POCKET GU

CATTLE BREEDS

OF NEW ZEALAND

GRAHAM MEADOWS

NH
NEW
HOLLAND

First published in 2008 by New Holland Publishers (NZ) Ltd
Auckland • Sydney • London • Cape Town

www.newhollandpublishers.co.nz

218 Lake Road, Northcote, Auckland 0627, New Zealand
Unit 1, 66 Gibbes Street, Chatswood, NSW 2067, Australia
86–88 Edgware Road, London W2 2EA, United Kingdom
80 McKenzie Street, Cape Town 8001, South Africa

Publishing manager: Matt Turner
Design: Amy Tansell / Words Alive
Editor: Alison Southby
Maps: Nick Keenleyside

National Library of New Zealand Cataloguing-in-Publication Data

Meadows, Graham.
Pocket guide to the cattle breeds of New Zealand / Graham
William Meadows.
Includes index.
ISBN 978-1-86966-226-4
1. Cattle—New Zealand—Identification. 2. Cattle—New Zealand.
I. Title.
636.20993—dc 22

10 9 8 7 6 5 4 3 2 1

Colour reproduction by Pica Digital Pte Ltd, Singapore
Printed by Times Offset (M) Sdn Bhd, Malaysia, on paper sourced from sustainable forests.

Contents

Preface

This book is written for people who have an interest in cattle and want a visually attractive reference book to add to their library. I hope it will appeal to the general public and owners of lifestyle blocks as well as to cattle farmers.

Like its predecessor, first published 12 years ago, the *Pocket Guide to Cattle Breeds of New Zealand* is intended to give an impartial and up-to-date summary of the breeds featured. This new book includes many previously unpublished photographs.

My thanks to all those who submitted information about their breed, and especially to Michael Trotter of the Rare Breeds Conservation Society of New Zealand, who supplied text about the Society and the image of the Enderby Island cattle.

Graham Meadows

PART ONE: Introduction

The evolution of cattle species

The common ancestor of modern cattle is thought to have evolved two to seven million years ago, during the Pliocene epoch. During the following Pleistocene epoch (two million years ago to about 11,000 years ago), notable for its series of Ice Ages, there evolved various members of the cattle family (family Bovidae). These include the Asiatic wild ox (genus *Bibos*), bison, buffalo, musk ox, yak and the primitive wild ox (*Bos primigenius*), better known as the aurochs or urus, which is the most likely ancestor of all of today's domestic cattle breeds.

The aurochs appears to have originated on the Indian sub-continent in and around the region that is now the Punjab. Remains of most of its various forms have been recovered from the hills of the Siwalik Range on the southern edge of the Himalayas. Originally native to the warm area extending from Turkestan in the north to the Indian and Arabian deserts, the aurochs gradually spread into Europe, but receded again with each successive Ice Age. It was not until the end of the Great Ice Age, about 250,000 years ago, that it spread permanently westward to the Middle East, northern Africa and Europe, and eastward into China. As the aurochs spread, the species

---

Aurochs

The word aurochs is derived from the Greek and Latin words for ox, *ouros* and *urus* respectively. The Old High German name was *or-ohso* which became *auerochs* in German and finally 'aurochs' in English.

Above: The aurochs was taller than even the largest modern cattle breeds, and also taller than the average modern man.

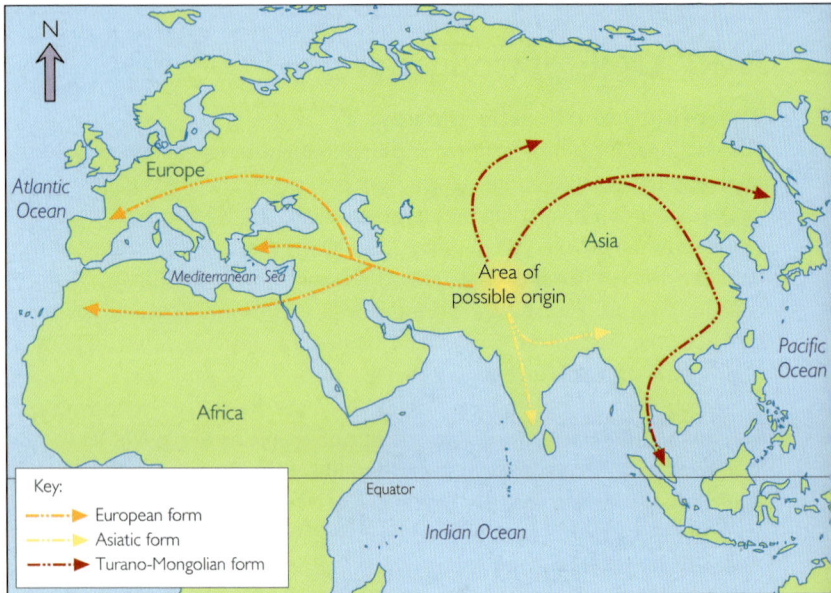

Above: The origin and spread of wild cattle.

began to differentiate and most authorities now recognise two major forms: the European form, *Bos primigenius primigenius*, from which evolved the domestic taurine (humpless) cattle (*Bos taurus taurus*); and the Asiatic form, *Bos primigenius namidicus*, from which evolved the domestic zebu (humped) cattle (*Bos taurus indicus*).

Archaeological finds suggest that the aurochs was a large, sturdy species. Bulls measured about 190 cm at the withers, about 38 cm taller than large modern breeds such as the Simmental and Charolais. Their horns were about 1 metre long, and they had massive necks and deep heavy forequarters. Aurochs cows were about 30 cm smaller than the bulls. Rock paintings in the caves of Lascaux in south-west France (dated to 13,000 BC) and at Jabbaren on the Tassili plateau in Algeria portray aurochs bulls as being darker in colour than the cows, ranging from dark red-brown to black, with a lighter coloured stripe along the spine. Most cows were coloured red-brown, as were their calves. White animals are also depicted, some with black speckles on the head and front of the body. Rock art portraying similar animals, unearthed in the Coa Valley in Portugal and first given worldwide publicity in 1995, is estimated to be about 20,000 years old.

Scientific names

The Swedish botanist Carl Linnaeus (1707–1778) first devised the system of using unique combinations of two Latin words (binomial classification) to describe a particular species. He used it in 1737 for a book on plants, and later devised a similar system for animals. Subsequent additions and alterations by other scientists have resulted in major changes to many of the early scientific names, but one that is still used today is the Latin word *Bos*, applicable to the genus of cattle from which domesticated species arose.

The scientific name for the aurochs, *Bos primigenius*, is derived from the words *Bos* = ox, and *primigenius* = first begotten. The European wild ox that was domesticated in northern Europe during the first millennium AD, the Roman 'urus', is still referred to in some literature as *Bos urus*.

The scientific name *Bos taurus* has been in use for more than a century and is still commonly used to classify humpless domestic cattle, while the name *Bos indicus* has been used for humped cattle. Because some breeds of domestic cattle are a mixture of the two, a more recent classification suggests the addition of a third word to separate the various forms. Terms used in this book are *Bos taurus taurus* or *Bos t. taurus* for humpless taurine cattle, *Bos taurus indicus* or *Bos t. indicus* for chest-humped (thoracic-humped) zebu cattle, and *Bos taurus africanus* or *Bos t. africanus* for the neck-humped (cervico-thoracic-humped) Sanga cattle.

The scientific name of zebu cattle, *Bos indicus*, is now deemed invalid by the *Integrated Taxonomic Information System (ITIS)*, which classifies the zebu under *Bos primigenius* along with all other domestic cattle and their extinct aurochs ancestors. *ITIS* is a partnership designed to provide consistent and reliable information on the taxonomy of biological species. *ITIS* was originally formed in 1996 as an interagency group within the United States federal government, involving agencies from the Department of Commerce, Department of the Interior, Department of Agriculture and the Smithsonian Institution.

Above: An artist's impression of cave paintings at Lascaux. Paintings found there and elsewhere clearly depict the aurochs as a massive beast, with a black and red coat. Other paintings show hunting scenes.

Above: A modern bison, the animal which most closely resembles the ancient aurochs. Thousands of years ago, these animals crossed the land bridge between Europe and North America, and because the population became isolated from other cattle, it remained relatively unchanged.

The domestication of cattle

By the beginning of the Holocene (current) epoch 11,000 years ago, cattle had already spread throughout Europe, Asia and Africa as a result of natural migration. One cattle type, the bison, had crossed the ancient land bridge into North America during the Pleistocene epoch, but had never progressed to South America. There were no cattle in Australia or New Zealand.

The world was already becoming much warmer. The ice sheets were retreating, and in south-east Europe and south-west Asia Stone Age people were starting to replace nomadic hunting with local, specialised methods, and creating permanent settlements on the plains or in the fertile and sheltered river valleys. It was around these settlements that the process of domestication of various species began.

Archaeologists have concluded that domestication of the aurochs occurred early in the Stone Age (9000–3500 BC), and followed that of sheep, goats, pigs and dogs. It probably started in a number of centres independently and at different times between 7000 and 3500 BC (late in the Mesolithic period to the end of the Neolithic period).

At present the earliest archaeological finds of domesticated cattle are from Greece and Anatolia (southern Turkey), both dated about 6500 BC, suggesting that there may have been one large centre of cattle domestication in the north-eastern basin of the Mediterranean. At around the same time, primary centres of domestication were also developed in three of the most important civilisations of that time: those of the valley of the Tigris and Euphrates in Mesopotamia (now Iraq), the Indus valley in India, and the Nile valley of Egypt.

In Mesopotamia the remains of domesticated cattle dated at about 6000 BC were found in the Zagros Mountains which lie on the north-eastern side of the valley of the Tigris and Euphrates (now part of Iran). To the north, at Diyala, other remains have been dated at about 5000 BC. Around the same time, cattle had also been domesticated in the Indus valley, and there is

Hamitic cattle

The word 'Hamitic' comes from the name for a group of African peoples whose members are called the Hamites. Ham was the second son of Noah, and the word Hamite is defined as 'a descendant of Ham'. Hamitic peoples speak related languages, are regarded as being of kindred origin, and include the Berbers of North Africa, the Tuaregs of the Sudan, and the Galla of East Africa.

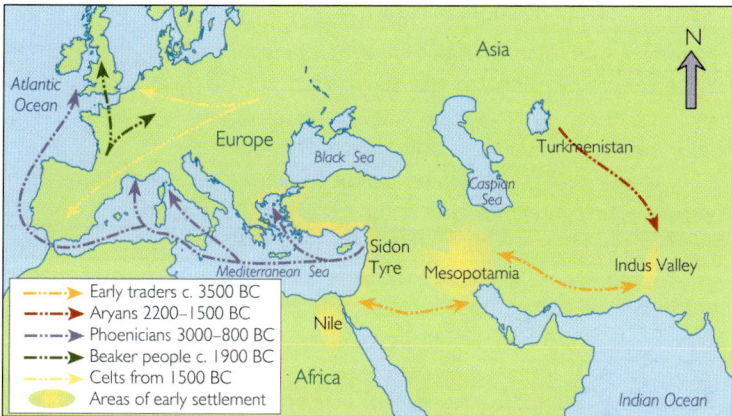

Above: The spread of domesticated cattle followed the traditional trading routes of their owners; thus, patterns of cattle spread strongly resemble the spread of early civilisations in Asia, Europe and Africa.

evidence that traders from the region had started to conduct a flourishing trade in humped and humpless cattle with their counterparts in Mesopotamia as early as 4500 BC. Indian and Mesopotamian traders also traded with the Hamitic people who were residents of the Nile valley in Egypt, where cattle were depicted on wall paintings from about 5000 BC.

Throughout the whole Middle East region, trade and migration resulted in the intermingling of humped and humpless cattle and the development of many different cattle types, both long-horned and short-horned. Adding to this movement and mixing of cattle were the Phoenician traders, operating mainly from the ports of Tyre and Sidon in what is now Lebanon. They were already trading throughout the Mediterranean by 3500 BC, and are thought to have reached Britain many centuries later, possibly trading cattle for Cornish tin.

Archaeological finds indicate that by early in the third millennium BC a comparatively sophisticated agricultural society was already in operation in Mesopotamia. Draught oxen were pulling ploughs and carts, butter and cheese were being manufactured, and cattle destined for beef production were being grain-fed during much of the year.

Two particularly interesting archaeological finds from the region are both dated *c.* 2400 BC. One is a mosaic frieze from a temple in the ancient Sumerian capital city of Ur, which was destroyed *c.* 2009 BC, depicting a cow being milked from behind, in the same manner as goats. The other is a picture on a fragment of vase from Tell Agrab showing a humped bull standing in a stall.

Centres of domestication also occurred in Europe. There is evidence of ploughing in the lower Danube valley (*c.* 4500 BC), in England and Poland (*c.* 3500 BC) and in southern Spain (before 3000 BC).

Secondary centres of domestication gradually developed throughout the European continent. At that time the climate was hot and dry, and huge herds of aurochs roamed the plains of Europe. One major centre was on the Hungarian Plain where the aurochs became the most commonly hunted animal, with the associated capture and domestication of its calves. Excavations of sites inhabited by Neolithic people have revealed large numbers of adult aurochs bones, from which it is inferred that the protecting adult cattle were killed to allow capture of the young calves for domestication. Other large centres of domestication were on the Wallachian Plain in southern Romania, and on the smaller and larger plains of Central and Western Europe. A distinctly separate centre of domestication later developed in Denmark (*c.* 2600 BC), with cattle from it spreading north and east.

Above: A depiction of milking from 5th Dynasty Egypt, *c.* 2350 BC.

Cattle types

Initially most cattle were of the *Bos primigenius* type, characterised by a flat forehead and long heavy horns. But along with domestication came the development of a number of distinct cattle types, which are of particular interest for their ancestral relationship to modern domestic cattle breeds. Because of a lack of historical records and other evidence their classification remains somewhat uncertain, but five major types are generally recognised. These are characterised by the absence or presence of a hump, the position of that hump, and the length of the horns.

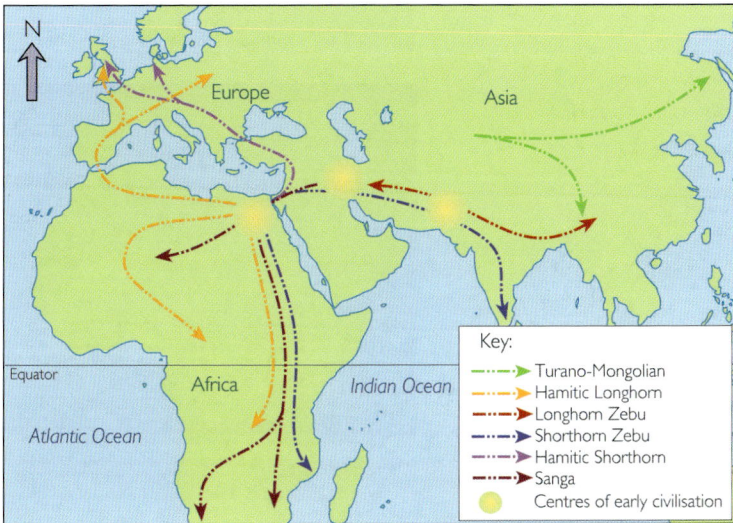

Above: The spread of major cattle types from their Middle-Eastern and North African origins.

Humpless cattle (*Bos taurus taurus*)

Hamitic Longhorn

The original cattle type of North Africa, depicted in Egyptian tomb paintings about 5000 BC, the Hamitic Longhorn was gradually displaced in two main directions by the Hamitic Shorthorn. To the south it was interbred with the Longhorn Zebu and gave rise to various Sanga types, including the Ankole of East Africa. To the south-west it gave rise to the Kuri of West Africa. To the west it entered Morocco, from where it was taken north into Spain. In around

Right: Ankole-Watusi cattle, descended from the Hamitic Longhorn, have long played an important role for many African tribes.

Right: Chillingham cattle, descendants of the Hamitic Longhorn, have remained remarkably unchanged for hundreds of years, and display many features associated with ancient cattle. Note the lyre-shaped horns.

3500 BC it entered Britain, where it eventually gave rise to the ancient Celtic cattle from which at least seven of today's British breeds were developed.

Hamitic Shorthorn

Depicted in Egyptian tomb paintings from about 2500 BC, the Hamitic Shorthorn gradually displaced the Hamitic Longhorn and spread westwards through North Africa to Morocco. Its main route into Europe was probably northwards through the Bosporus, gradually spreading into Switzerland, west past the Massif Central region of France, and north to Brittany and the Channel Islands. It entered Britain around 1500 BC. The modern breeds of the Mediterranean region are mainly of this type, and breeds that may be descended from it include the Brown Swiss, Aubrac, Blonde d'Aquitaine and Jersey.

Left: The Aubrac, a likely descendant of the Hamitic Shorthorn.

Humped cattle

The exact origin of humped cattle remains unclear. Artwork on friezes and pottery indicates that domesticated humped cattle were originally in Mesopotamia (now Iraq) by 4500 BC. The oldest skeletal remains of humped cattle, dating from 3200 BC, were found on the edge of the Indian desert. These were undoubtedly domesticated, and to date no remains of wild humped cattle have ever been found. Because of this, many authorities maintain that humped cattle arose through human selection following, not prior to, domestication, and that because the two forms can be interbred they should be treated as one species. However, they remain classified separately because of their significantly different biochemical, histological, anatomical and digestive features.

Humped cattle are now generally classified into two types according to the position of the hump: zebu and Sanga.

Zebu cattle (*Bos taurus indicus*)

In zebu cattle the hump lies over the shoulders (chest-humped or thoracic-humped). Its shape varies from breed to breed: it may be round or pyramidal, smooth or irregular, or any intermediate shape between these. It is composed mainly of muscle tissue, and its function is unknown. Zebu cattle also have very loose skin, which hangs in folds along the underside of the neck forming a dewlap that may be large or small according to breed. The resulting additional skin area may enable the animals to radiate heat more efficiently, and more effectively regulate body temperature in very hot conditions.

Other zebu features include: a long head; oblong eyes; large, often pendulous ears; a quick, light gait in which the hind legs are carried forward in a straight line; and a characteristic voice.

Longhorn Zebu

A chest-humped type with characteristic lateral horns which was recorded in the old civilisations north of the Persian Gulf, the Longhorn Zebu was probably spread throughout India by the Rig-Vedic Aryans who travelled from the north-west through the Khyber Pass into the Indus valley between 2200 and 1500 BC. It eventually spread throughout south-east Asia, where it became intermixed with humpless cattle derived from a distinctive prehistoric cattle type known as Turano-Mongolian, which entered from the north via China.

Right: The Africander, a Sanga type, possibly the closest living relative of the extinct Longhorn Zebu.

Right: The Red Sindhi, a classic example of the Shorthorn Zebu.

The Longhorn Zebu also spread westwards as far as Azerbaijan, and may have entered the Horn of Africa with migrating Semitic tribes through what is now Ethiopia between 2000 and 1500 BC, spreading across that portion of Africa that lies below the Sahara desert, north into Egypt and south towards the Cape of Good Hope. Crossing with the Hamitic Shorthorn resulted in the Sanga cattle type. The Longhorn Zebu is now extinct, its closest existing relative possibly being the Africander.

Shorthorn Zebu
A chest-humped type, probably developed about 2000 years ago, the Shorthorn Zebu formed the basis of modern Indian breeds. It was also taken down the east coast of Africa by Arab and Indian traders from about AD 650, and is now the dominant type in East and Central Africa.

Sanga cattle (*Bos taurus africanus*)
Sanga cattle provide yet another puzzle. They have a slender conformation with long legs, long to very long horns, and a hump that lies forward over the base of the neck (neck-humped or cervico-thoracic-humped). Some

Above: A Tuli bull, showing the forward-lying hump that characterises Sanga breeds.

Current distribution of humped and humpless cattle

As a rule of thumb, tropical environments are stocked mainly with humped cattle, and temperate or cool environments are stocked mainly with humpless cattle.

Humped cattle. In Asia humped cattle are found from southern Arabia and southern Iraq through Iran, Afghanistan, the Indian sub-continent and south-east Asia to southern and central China, and in parts of southern Russia. In Africa they are ubiquitous south of the Sahara. In Australia just over 20 per cent of registered cattle have varying shapes and sizes of hump depending on their composition. Some of these are pure-bred African and Asian cattle breeds, while others have been developed from crosses between various *Bos t. taurus* and *Bos t. indicus* breeds to provide heat and drought resistance.

Humpless cattle. In Britain and Europe all the native breeds are humpless. In Asia humpless cattle prevail north of humped cattle territory, being found from Israel in the west through northern Iraq and Iran, the whole of Tibet and Mongolia, Manchuria, Korea and to Japan in the east. They occur through the major part of Soviet Asia as far north as cattle are kept. In Africa there are native humpless cattle in Egypt, Algeria, Morocco and parts of West Africa. Most are short-horned, a very few are long-horned. In southern Africa there are many humpless breeds of European derivation. In Australia almost 80 per cent of the cattle on farms are of European origin and humpless, while in New Zealand that figure is closer to 100 per cent.

authorities think they may have originated about 1600 BC in the region that is now Ethiopia and Somalia, through the interbreeding of resident Hamitic Longhorn cattle and Longhorn Zebu cattle that had arrived some 400 years earlier. Supporting this hypothesis is the fact that cross-breeding a humpless animal with a chest-humped animal usually results in neck-humped offspring. But in Libya and the Ahaggar mountains of the Sahara, rock paintings estimated to be nearly 8000 years old depict Sanga-like longhorned cattle, and it has been suggested that Sanga animals might have arisen as an original strain of humped cattle independent of any Asiatic zebu influence.

Whatever their origin, Sanga cattle eventually spread southward through Africa along with nomadic tribespeople, leaving behind pockets of pure Sanga breeds and various zebu–Sanga crosses. Sanga cattle are now widely spread in South, Central and West Equatorial Africa. The indigenous breeds of Zimbabwe all fall into this group.

Factors contributing to the development of cattle types and breeds

Very slow change. Occurs over many millennia

1. Ancient migration of cattle, adaptation and change in new environments. Examples: aurochs and Asiatic wild ox.
2. Domestication of wild cattle by humans and movement of cattle between populations. Example: Longhorn Zebu.
3. Development of distinct types of cattle in specific geographical areas. Examples: Boran in southern Ethiopia, Limousin in France, Red Sindhi in Pakistan, Chillingham in Britain.
4. Migration and mixing of different cattle types to produce new local types. Example: ancestors of Piemontese in Italy.

Gradual change. Occurs over many centuries or during a few millennia

1. Primitive human selection of native cattle types for specific characteristics and/or uses, and ongoing breeding to retain these. Example: Sahiwal of India and Pakistan.

Moderately fast change. Occurs over many decades

1. Breeder selection and development of modern cattle breeds. Examples: Aberdeen-Angus and Hereford in 18th and 19th century England, Murray Grey and Droughtmaster in 20th century Australia.

Fast change. Occurs within a decade

1. 'Scientific' development in a country of new breeds to suit particular environments or climates. Example: Belmont Red in Australia.

Rapid change. Occurs within months

1. Export/import of cattle to/from countries around the world by means of live animals, artificial insemination or embryo transplant techniques. Examples: Hereford imported into Australia and New Zealand by early settlers, novel breeds established in various countries through embryo transfer and artificial insemination.

Note: Changes are influenced by climate and environment.

Africander cows and calves

Boran bull

Sahiwal cow

Droughtmaster

Belmont Red bull

Belgian Blue calf with recipient cow

The uses for cattle

Until comparatively recently cattle had a threefold purpose: meat, milk and draught. Of course certain animals performed better in a particular respect than others. For example, in the 13th century Gloucester cattle were already noted for the cheese made from their milk, and the Charolais was highly rated for its meat early in the 17th century. But it was not until about 200 years ago that farmers started to make conscious efforts to develop specific 'culture' breeds for either beef or milk production.

Meat

It is almost certain that the main reason for domesticating and keeping cattle was to eat their meat, although some were kept purely for sacrificial purposes. Early Stone Age art also endows cattle with some magical significance, and by the Late Stone Age the bull had become a fertility symbol.

Nevertheless, archaeological evidence suggests that throughout the Neolithic period cattle accounted for 40–90 per cent of all domestic animal meat eaten, the proportion depending on the type of husbandry in an area and the availability of other food species such as sheep, goats, pigs and dogs. The use of cattle for meat continued throughout the centuries and early into the 19th century, from which time specific beef breeds were gradually developed (see Cattle breeds in Britain, page 27).

Above: (*left*) Gloucester cows have been famous for their milk since the 13th century. (*right*) The Hereford is a popular beef breed.

Milk

Milk production would initially have been minimal, for cows on a low level of domestication would have had little or no milk to spare after feeding their calves; however, the death of a calf would have provided humans with an opportunity to milk the mother. But as humans learned better husbandry and introduced grain-feeding, better selection for milk production became possible.

The first representation of milking taking place originates from southeast Iraq in the temple of Ninhursag in Ur, which was once the capital of the ancient Sumerian Empire, and is dated around 2400 BC, although it is logical to assume that cows were milked long before then. It is interesting to note that the cows are shown being milked from behind, as is usual with goats. In classical Greece and Rome, milk production was very important. Aristotle refers to cows of the large Molossos breed giving one amphora of milk a day. It is said that breakfast for the average Roman was cow's milk, while colostrum milk and whipped cream were considered delicacies.

Two millennia later milk is still a staple food in many countries, the biggest differences being the development of specific milking breeds and the methods and volume of production.

Above: The Holstein-Friesian is now the most popular milk breed in New Zealand.

Draught

Cattle were already being used for draught purposes in Mesopotamia during the Neolithic period, but their draught power was particularly recognised by the Greeks and the Romans. An ox could not move as fast as a draught horse but it could pull a far heavier load, and the use of oxen for draught purposes gradually spread.

By the 18th century oxen were an integral part of most farming operations. The early settlers in Australia and New Zealand used teams of oxen for bush and forest clearance, and they proved ideal for hauling timber out of the densely forested areas. In many countries oxen were eventually replaced by heavy horses such as Shires and Clydesdales, which in their turn succumbed to the introduction of mechanisation, and in most of Europe and Australasia the only places one is likely to see draught oxen are at agricultural shows or farm parks. It is a different story in Africa and Asia, where draught oxen still play a very important role in agriculture, especially in developing countries.

Above: A woodcut of mediaeval ploughing with oxen. Source: *Mediaeval and Modern History* by Philip Van Ness Myers, 1905

Right: A modern-day farmer ploughing with oxen in Ethiopia. In many countries agricultural techniques have changed little from those used thousands of years ago. Photograph courtesy Keith and Carolyn Hamill.

The development of cattle types in Britain

There is evidence that cattle were already domesticated and being used for ploughing in Britain by 3500 BC. The process of domestication was continued first by the Beaker people, who arrived around 1900 BC, and later by the Celts, who invaded in 750 BC. The cattle that the latter possessed, now referred to as Celtic cattle, were a mixture of the original indigenous cattle and the descendants of the Hamitic Longhorn cattle of Egypt.

When the Romans invaded and occupied England, the Celts retreated to the south-west (Devon and Cornwall), the west (Wales), and to the north (north-eastern England and Scotland), taking many of their cattle with them. From these Celtic (ancient Hamitic) cattle are derived the Highland, Welsh Black, Chillingham, White Park, Kerry, Dexter and Vaynol.

The Romans brought in their longhorned red cattle, mainly for use as draught oxen, and as time went by the Celts acquired some of them. After the fall of the Roman Empire in AD 476 the retreating Romans left many of their cattle behind and these became assimilated within the indigenous cattle population.

The invading German Angles arrived in AD 500, and the Saxons arrived from AD 600 onwards, gradually intermingling with the Angles. These Anglo-Saxons introduced more cattle from Europe.

Between the 7th and 10th centuries the Vikings were raiding and invading, introducing red cattle into Ireland between the 8th and 9th centuries (giving rise to the Irish Moiled), and also into Normandy and the east of Britain (leading to the Suffolk Dun from which was derived the old Norfolk Red and the Lincoln Red).

Left: Welsh Black cattle are descended from the ancient Celtic cattle.

Above: The Lincoln Red was developed from cattle introduced into Britain by the Vikings.

After the Normans defeated the Anglo-Saxons at the battle of Hastings in 1066, they introduced the red cattle of Normandy, the ancestors of the Devon, South Devon and possibly the Sussex.

Throughout this period, however, the standard of cattle husbandry declined badly, and by the early Middle Ages (10th to 13th centuries) the withers height had decreased to a new low of just over 1 metre. It increased again only after the Renaissance period (14th to 16th centuries), when conscious breeding was once more introduced and new cattle varieties were developed.

But for the wild ancestor of domesticated cattle, the aurochs (*Bos primigenius primigenius*), it was the end of the line. It was already extinct in western Europe by the 15th century, and the last known survivor, a cow, was killed in 1627 by poachers in a hunting reserve near Warsaw, Poland.

The aurochs is still possibly represented by three races of modern domestic cattle – the Spanish, Camargue and Corsican – but of the two original wild ox types only the Asiatic wild ox remains, represented by such species as the Gaur (*Bibos gaurus*), Gayal (*Bibos frontalis*) and Banteng (*Bibos sondaicus*).

Above: Chillingham cattle.

Cattle in Britain – chronology

BC

9000	Wild cattle (aurochs) are already in Britain at the start of the Holocene period.
3500	Cattle of Hamitic Longhorn type have already entered Britain. Evidence suggests that they have been domesticated and are already used for ploughing.
3000–800	Phoenician traders may have introduced some cattle to Ireland and southern England.
1900	Beaker people spread from Spain and central Europe into Britain.
1500	Cattle of Hamitic Shorthorn type are introduced.
750	The Celts invade Britain and acquire British native cattle, which by now are of mixed ancestry but mainly derived from the Hamitic Longhorn and Shorthorn types.
700	Ancient Celtic cattle well established.

AD

55	The Romans invade Britain and start to introduce their longhorned red cattle. The Celts retreat to the 'Celtic fringes' of south-west England, Wales and Scotland, taking many of their cattle with them.
500	The Angles enter Britain and introduce additional cattle from Europe.
500–600	The Saxons enter Britain. More European cattle introduced.
700–800	The Vikings introduce cattle from Scandinavia into Ireland and the east of England.
1066 onward	The Normans introduce the red cattle of Normandy.

The development of cattle breeds in Britain and Europe

For centuries cattle slowly developed as types rather than breeds in the modern sense. There was significant intermixing, and few records were kept, with the result that by the 18th century the ancestry of most cattle was unknown.

There were exceptions. For example, by the 12th century cattle on the comparatively isolated mainland and islands of Scotland were already of a distinctive type, now called the Highland. By the end of the 13th century Gloucester cattle were recognised for their special type of milk, which could be made into high-quality cheese, and the wild cattle now known as the Chillingham herd had been enclosed within a walled park that kept them almost totally isolated.

Above: South Devon heifers, descended from the red cattle of Normandy.

In the 17th century the Benedictine monks at the Abbey of Aubrac in the south of France organised a rationally managed herd, and from their records the history of the Aubrac breed can be traced back at least that far. Around the same time the indigenous cattle of the Simmen Valley in Switzerland, from which the modern Simmental evolved, were already famous for their draught power, and their meat and milk production. The cattle on the Channel Islands of Guernsey and Jersey, developed specifically for milk and cheese production, had also remained comparatively isolated and retained many of the characteristics of their ancestors.

Conscious breeding

The techniques of conscious breeding were initially developed in south-west Asia, and spread to the Mediterranean. Conscious breeding implies the use of basic knowledge to breed for desired characteristics, based on the premise that 'like produces like'; by 500 BC this knowledge was already being developed by the Greeks and the Scythians. The Scythians were a nomadic Indo-European people who settled in an area north of the Black Sea which became Scythia, and among their cattle herds hornless individuals were common.

But it was the Romans who made the most significant changes. With the first introduction of conscious animal breeding in the Roman Imperial Period (which commenced in 31 BC) the size of cattle rapidly increased, eventually averaging 125 cm but reaching 140 cm or more. The Romans developed a large longhorned cattle breed which was introduced into the provinces of the Empire all over Europe, but opinion is divided as to what influence it had on the aboriginal cattle populations.

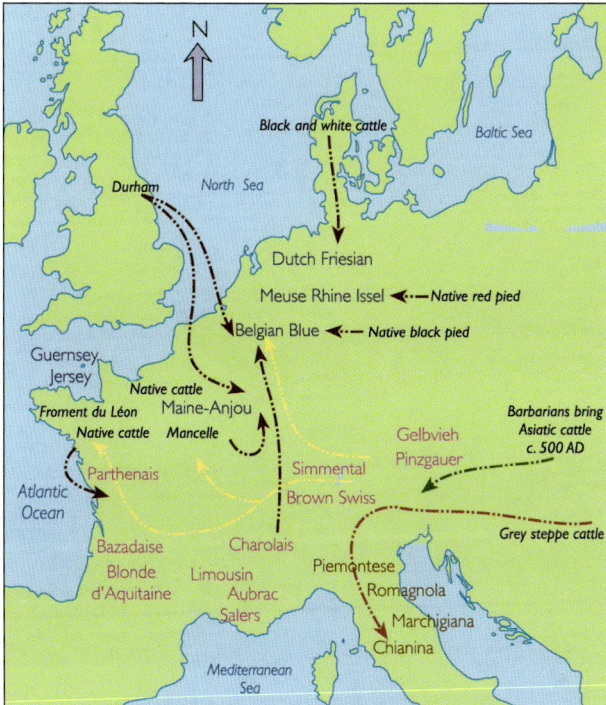

Left: The spread of cattle breeds into Europe and Britain.

The introduction of controlled breeding

Although early examples of breeds were already in existence, Robert Bakewell, an English agriculturalist (1725–1795), is generally credited with making breed development possible by introducing a new idea into the age-old principle of like-breeds-like: the systematic use of controlled inbreeding.

Using this method he developed native longhorned cattle from central England into an improved breed, the Dishley or Leicester Longhorn. Other breeders followed suit, among them the Collings brothers, who developed shorthorned cattle from northern England into the Durham, the pioneer breed for the Shorthorn that soon became the most popular English breed.

For this new system of breeding to be successful it was essential to keep accurate records. It was also important for breeders to keep in touch and help promote their selected breed. So eventually, many decades (or in some cases more than a century) later, breed societies were formed. From these came herd books, in which detailed information on pedigree breeding stock was recorded. Of the breeds profiled in this book the earliest recorded herd book is that of the Simmental, which was produced in 1806. The first Coates's Herd Book for the Shorthorn was published in 1822.

Above: The Durham ox, the pioneer breed for the Shorthorn, as depicted in a contemporary print.

Cattle breeds in Britain

Alongside breed development came the selection of cattle for more specific purposes than the original roles of meat, milk and draught. Most breeds, such as the Shorthorn, were originally developed as dual-purpose animals, but before long more specific dairy and beef Shorthorn lines appeared. Further development also occurred with some of the more established single-purpose breeds, such as the Aberdeen-Angus, Hereford and Ayrshire.

The industrial revolution also had a significant effect on breed development and popularity. As more and more workers moved from the countryside and earned good wages in the towns, the city demand for meat and milk rose. Breeds that did not produce those commodities efficiently went out of favour; those that did became popular. For example, improvements in road and rail transport enabled Scottish farmers to send their Aberdeen-Angus cattle to English markets where they commanded a better price, so the breed became a mainstay of Scottish beef production.

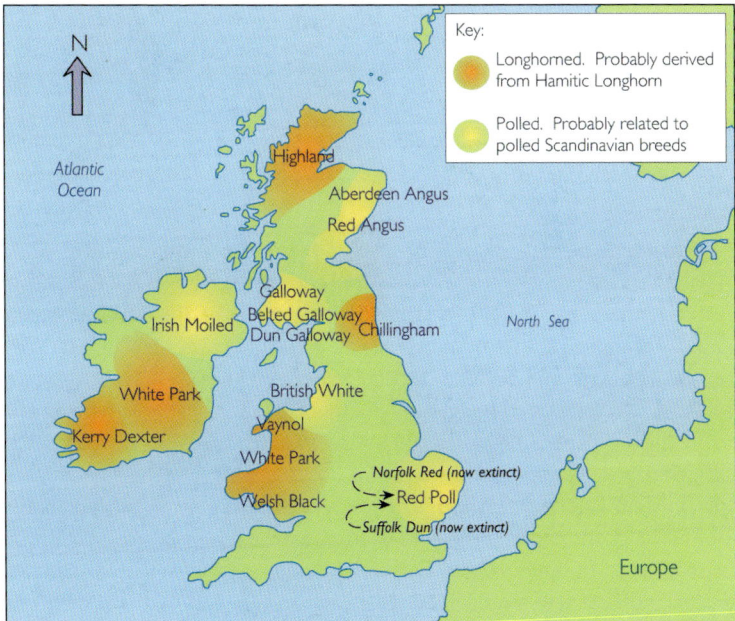

Above: The distribution of longhorned and polled cattle in Britain.

Right: The origins of British Lowland and solid-coloured breeds.

But although single-purpose breeds had their place, until the end of World War II the dual-purpose cow was the economic base on which the majority of British farmers survived, for beef from dairy herds was just as important to the nation's economy as that from beef herds. The dairy farmer had a product to sell in addition to milk and the cull dairy cow (boner beef) – namely, the dairy beef calf, which gave the beef farmer another type of animal to fatten and sell.

In the 1960s and 1970s the situation changed radically. The pursuit of single-purpose breeds and extremes in both dairying and beef resulted in the Holstein-Friesian becoming the premier dairy breed, and in previously 'unknown' Continental beef breeds such as the Charolais appearing on British farms. In the European Economic Community subsidised production created mountains of unsold butter that had a major effect on the agricultural economies of Australia and New Zealand. In 1974 the first major downturn occurred in the meat industry, eventually reducing the numbers of beef cattle and forcing rationalisation within the industry.

The physical characteristics of British breeds were also changing, with some beef types being selected for longer, larger bodies and longer legs to produce a higher yielding carcass; a process that sometimes involved the use of a Continental breed to 'upgrade' the existing British breed. One example of this was the use of the Maine-Anjou to improve the Beef Shorthorn. Consumer demands also changed: more people became vegetarians or vegans, while many of those who ate meat demanded less fat and lower cholesterol levels from leaner carcasses. Lean meat without fat was tasteless, so selection occurred to produce animals that had fat evenly distributed between the muscle fibres (marbling) to give better flavour.

The variety of breeds that make up the British national herd is still changing, and development of many of the established British breeds continues. Experience now suggests that breeding for increased growth comes with some undesirable consequences: a heavier birth weight brings associated calving difficulties (in some breeds, such as the Belgian Blue, almost all deliveries are veterinarian-assisted), and delays in sexual maturity lead to lower heifer fertility; so the final result may be no net improvement in a herd's productivity.

The majority of today's popular breeds are single-purpose: either beef or dairy. Some regard this as beneficial, others see it as a retrograde step; but the fact remains that there are very few true dual-purpose breeds left. Of the breeds featured in this book, those that best fit the dual-purpose description are Brown Swiss, Dairy Shorthorn, Dexter, and perhaps the Meuse Rhine Issel.

Below: (*left*) Scottish beef breed Aberdeen-Angus. (*right*): The Brown Swiss, a dual-purpose breed.

Cattle breeds in New Zealand

During the 19th century the breeds of cattle farmed in New Zealand were almost exclusively of British and Channel Island origin. But with the advent of refrigerated shipping many things changed. The nation became more and more dependent on the export of meat and dairy products to the United Kingdom, and its cattle farmers sought maximum production from as much available land as possible.

Because of the country's largely warm temperate climate there was little need for New Zealand farmers to change their existing breeds. The Hereford and Aberdeen-Angus became the prime beef producers, and first the Shorthorn and then the Jersey the principal dairy breed. Some beef production came from dairy herds, through the mating of beef bulls with those dairy cows not required to produce calves for herd replacement, but the accent was still very much on single-purpose breeds. Because the available British or Continental breeds filled the bill, there was no necessity to develop a unique New Zealand cattle breed.

As elsewhere, the popularity of different breeds has varied over the years. The most significant change has been the rise of the Holstein-Friesian, which has now displaced the Jersey as the number one dairy breed. Some Continental beef breeds, such as the Charolais and Simmental, have become well established; others that are less well known are being introduced. In general the breeds in New Zealand have remained fairly constant, and the most significant changes have been in their numbers.

Genetics

Because of the intermixing of cattle types and the lack of early records it is difficult to determine the ancestral relationships between many of today's breeds. A number of different theories have been put forward, based on information from one or more of three sources: the breed's phenotype (what it looks like); its genotype (genetic make-up); and historical evidence.

Phenotype

The phenotype includes key morphological characteristics such as the shape and size of the head and horns, which are likely to be less affected by selection. The predominant coat colour of a breed may also provide a clue, although this can be misleading because it may have been influenced by human selection. For example, to the Celts red animals were a symbol of fertility and crops, while black ones symbolised pestilence and death. Body shape and size are likely to be less useful because they are influenced by the type of management and/or the environment.

1930   1960   1990

Above: The shape of the Aberdeen-Angus has changed during the last hundred years as tastes in meat have altered.

Right: Devonshire cattle in an 1808 print.

Left: Beef cattle near Muriwai, north of Auckland.

Devonshire Cattle

Genotype

Biochemical evidence obtained through blood-typing, genetic finger-printing and other forms of DNA analysis has also proved useful, although its accuracy depends on the 'purity' of the cattle tested. It may be misleading where a breed has suffered major introgression through the introduction of other bloodlines. For example, in the 19th century Devon cattle, and possibly Highland cattle, were used to improve the Salers, significantly changing its genetic profile.

Historical evidence

Archaeological sites have yielded cattle bones that give a general idea of the cattle types of particular periods, although in the past much of the evidence has been misleading. Ancient friezes, murals, and cave or tomb paintings have proved more useful. Particularly valuable have been ethnological studies of human movement and migration, especially when related to trade or settlement.

Changes within breeds

Morphologically and genetically few, if any, breeds remain static. Fashions of the time dictate the shape and size of show animals. Market forces also dictate change. For example, consumer resistance to fatty meat has necessitated selection to reduce the depth of backfat and improve marbling in the meat. Selected to provide a market for 'baby beef' during the 1940s and 1950s, the Beef Shorthorn, Britain's most popular beef breed in the 19th century, became a victim of market change and is now classified as a rare breed. Changes can also occur when another breed is used to create improvements in the original (introgression). For example, in the early 19th century French breeders imported Shorthorned Durham cattle (predecessors of the Shorthorn) to improve the Mancelle breed and create the Maine-Anjou. More recently, in an interesting reversal, the Maine-Anjou has been used to develop the improved Beef Shorthorn.

Right: The Maine Anjou was used to develop the improved Beef Shorthorn.

Changes in breed popularity

Breeds, like fashions, come and go. But since the 1970s there have been two significant changes which have affected the cattle industries of Britain, Australia and New Zealand. The first is the rise in popularity of the Holstein-Friesian, which has become the most numerous dairy breed in response to a desire to produce high volumes of milk with a high protein–fat ratio. The second change has been the rise in popularity of certain large Continental breeds, initially involving the Charolais in the 1960s but now encompassing many other breeds including the Limousin and Simmental. Many farms have introduced or are experimenting with Continental breeds, either for cross-breeding or for the establishment of small satellite herds.

Two other factors are having an impact on the prevalence of breeds. The first is the technique for multiple ovulation and embryo transfer (MOET), which although costly can prove beneficial to farmers wanting to rapidly upgrade their stock or establish a novel breed.

The second is the establishment of the Rare Breeds Survival Trust in the United Kingdom, and the Rare Breeds Society of New Zealand, reflecting the increasing interest being shown in the less common breeds of cattle (and, incidentally, other species), by hobby farmers and small-block owners.

Above: The Limousin is just one of the many Continental breeds to enter New Zealand.

Left: A South Devon calf born to a Black Simmental cow, the result of embryo transfer.

Cattle genes and human influence

As local types of cattle evolved and adapted to their environment they developed particular genetic profiles to enable them to survive. Cattle in very hot climates developed pigmentation of the skin and mucous membranes, and hooded eyes to help them to cope with the sun's strong ultraviolet rays. They developed loose folded skin in a dewlap to help them to maintain an even body temperature.

Those in drought-prone areas developed foraging ability and strongly built hooves to cover the hard rough ground. They developed resistance to the attacks of biting insects such as ticks, the depredations of internal parasites, and had a reduced incidence of bloat. Cattle in wet marshlands developed wide splayed hooves for easier movement. Those in cold or mountainous regions developed a tolerance to low temperatures and icy winds, and grew thick hair coats.

By the time the Greeks and Romans were developing particular cattle types there had already been millennia of subtle human selection pressures, and in many areas the genetic make-up of domesticated cattle was significantly different to that of their wild counterparts.

Although cattle breeding reached a low point during the Middle Ages, much of the gene pool remained. From the 18th century on, when 'conscious breeding' led to the development of the more modern 'culture' breeds, breeders started to exert a more significant influence on the genetic make-up of the animals in their care.

Left: The Brahman is adapted for hot climates.

Right: The Highland is a heavy-coated, cold-adapted breed.

The effects of human selection

By the time humans started to impose comparatively crude methods of selection on their domesticated cattle the latter were already the product of hundreds of thousands of years of genetic evolution. Initially human selection methods were based on simple objectives, but incidentally selected for particular genes. In those countries where humans used their cattle for meat or milk production as well as draught, owners tended to keep the offspring from the more placid animals (temperament); the cows and bulls that produced plenty of healthy calves (fecundity and fertility); the cows that gave the most milk (production); or those that produced the largest or strongest offspring (meat or draught). Animals that did not perform, or were particularly susceptible to the stresses and diseases of a domesticated environment, either died or were slaughtered.

One notable exception to this was the cattle kept by the Hindu people of the Indian continent, where cows were regarded as sacred and cattle were kept mainly for their draught ability. It has been suggested that historic lack of such selection in zebu (*Bos t. indicus*) cattle may partly explain the comparatively low fertility rate of Brahman when farmed under the tough tropical conditions of northern Australia. Their average weaning rate is about 50 per cent, compared with the 80 per cent obtained in temperate climates from the common European *Bos t. taurus* breeds, and although poor nutrition is clearly an important factor it is possible that inherited low fertility also plays a part.

The use of genetics in creating new breeds

Development of most of the breeds listed in this book started several hundred years ago, through a series of breeding programmes in which the most desirable characteristics of the more ancient breeds were combined to develop distinctively new breeds. The science of genetics was in its infancy, but advances since then have enabled modern breeders to select more rigorously and effectively for desired features, and spawned a completely new approach to cattle breeds and breeding.

The situation was changed with the introduction of breeding programmes based on heterosis, or hybrid vigour, where the most desirable traits of each parent tend to be expressed in their offspring: the F1 generation. But this success created another problem. The F1 generation was fine, but back-crossing with either of the original parents resulted in a loss of either resistance or production potential. Mating the F1 with another F1 to produce an F2 resulted in a resegregation of the genes controlling resistance and production potential, and the F2 lost some of the advantages conferred by either parent and often turned out be an intermediate performer. If a new breed was to be created through such a cross-breeding programme, it would take at least five generations, and by the end of that time the original advantages seen in the F1 generation could be lost.

Above: A cross-bred Maine-Anjou × Hereford calf, a classic example of hybrid vigour.

Genomic selection and sexed semen

Semen sexing technology has been available in a restricted sense for many years but has not been economically viable to implement in commercial farms. Sexed semen is now being produced and marketed by companies in Holland and Brazil. In New Zealand AmBreed (the second largest artificial breeding company in New Zealand) has been running a pilot trial in cooperating farmers' herds to test in-calf rates and efficiency of sexed semen from some high-ranking Holland Genetics bulls, with the aim of introducing Holland Genetics and AmBreed sexed semen in 2008.

Genomic selection will allow breeders to more accurately predict the genetic make-up of a particular young bull before the bull enters progeny testing. This is done by analysing the DNA of the young bull to identify its genetic make-up at tens of thousands of different positions on the chromosomes. The result will be a better estimate of breeding values for each young bull.

For more information about semen sexing, see the AmBreed website: www.ambreed.co.nz.

Heterosis is still a poorly understood phenomenon, and work on the subject continues. One of the avenues being explored is the use of a third, unrelated cattle type to breed with the F1.

Above: A Hereford × Friesian cow with a pure-bred Blonde d'Aquitaine calf – the result of embryo transplant.

The price of progress

During the last hundred years particular breeds of cattle have been selected in environmentally favourable areas to give very high production. Widely transported around the agricultural areas of the world, they displaced the lower-producing but more adapted local breeds. In some cases this was through direct physical substitution; in others through the use of grading-up programmes (see Genetics on the farm, page 44). This process was rapidly accelerated by the use of artificial insemination and embryo transplant, and emphasis was placed on breeding for high production based on recorded performance and the standardisation of animal products in marketing.

Our improved knowledge of genetics has led to dramatic changes. Breeds that did not shape up went to the wall, with the concomitant loss of their (possibly unidentified) genes. In the beef field, selection for fatness or leanness, rapid growth and a short life to slaughter has completely changed the appearance of some breeds within a few decades.

Broad-spectrum parasiticides have eliminated the need to select for parasite resistance, but this trend may need to be reversed with the emergence of drench-resistant parasites. Improved veterinary techniques have achieved better results from assisted calvings and Caesarean section has eliminated the need for animals to calve on their own, so oversized calves are viewed by some as less of a problem. This is human selection at its most potent, and the consequences are not yet fully understood.

Heterosis has been a major factor in the development of the dairy-beef industry. A Belgian Blue bull crossed with a Holstein-Friesian cow, or a Devon bull crossed with an Ayrshire cow, will usually produce F1 calves with the best attributes of both parents. There is no need to be concerned about these F1 calves' future breeding programmes because they are destined for slaughter. As long as pure-bred sires and dams remain available, the calves will keep coming.

One notable exception to this pattern is the Hereford × Friesian female, a particularly good performer that is frequently kept either for multiple-suckling of fostered calves, or, more recently, to be a recipient cow for embryo transplant.

This present era might well be described as the one in which the pursuit of heterosis reached its height. Read almost any beef breed literature, and you will find the bulls recommended for cross-breeding programmes as terminal sires. Cross-breeding is the buzz word, and heterosis or hybrid vigour is the reason.

Breed preservation

By the early 1970s a number of people had become concerned that the head-long march towards the extreme dairy or beef type cow was being made at the expense of good historical and market forces. They argued that breeding down narrow avenues of extreme types was not good for the long-term well-being of farmed cattle, and expressed concern at the ever-shrinking genetic pool. It was during this period that efforts were commenced to preserve the uncommon or rare British breeds, many of which appeared headed for extinction, and in 1973 the Rare Breeds Survival Trust was established in Britain (see page 41).

This action proved timely, for by the early 1980s Holstein-Friesian cattle had become favoured in almost all the lowland areas of the European continent and the British Isles, while in the moderately elevated areas of central and south-eastern Europe the change had been towards Simmental. Along with these changes had come the decline or disappearance of local native breeds. For example, by 1983 only 33 of 148 recognisable local breeds in Europe had held their own. The remainder were in danger of extinction.

The question asked was whether the extinction of native breeds was causing the loss of valuable genetic resources. Far too little was known about them. Nobody had any idea how they would perform if treated in the same way as 'improved' breeds, what commercial importance their adaptation to local areas could have, or whether using them for cross-breeding would result in hybrid vigour (see The use of genetics in creating new breeds, page 36). Some people were also concerned about the sociological aspect: many local breeds were part of an ethnically primitive form of animal husbandry, and as 'progress' destroyed the ethnic culture the local animals also disappeared.

Left: The Longhorn is an ancient breed, in which interest has now revived.

The conservation of genetic resources

If genetic resources were to be conserved, then people needed to decide what to conserve, and how. Perhaps breeds should be evaluated not only for their efficient production of desirable products, but also for their distinctive genetic characteristics (even if currently of no economic importance) in order to conserve as large a range of the existing, irreplaceable, genetic variation as possible. There were some who argued that only after such evaluation could an informed decision be made about a breed's future.

In the early 1980s a number of strategies were suggested for the preservation of genetic material. Some of them are described below.

Maintaining populations of each breed or strain. This is an expensive exercise that requires the establishment of a breeding programme that will avoid genetic change. Measures include an effective population size of perhaps 50 animals, breeding from older individuals, reducing inbreeding, a 1:1 ratio of breeding males to breeding females, and ensuring that each male is replaced by his son and each female by her daughter. At least two sizeable herds need to be kept, spaced well apart. Other small herds should be established wherever possible in places such as zoos, wildlife parks and recreational parks.

Setting up genetic stores. These include frozen semen, frozen fertilised ova and frozen gonadal tissue at several locations to guard against loss or technical failure at one site.

Conserving 'inferior' genetic stocks. These do not contain genes for high production or specific adaptations.

Formulating national breeding policies. These are based on performance recording so as to reduce selection differentials, thereby keeping a broad genetic base.

Right: A large semen and embryo storage unit operated by Xcell Breeding Services.

The Rare Breeds Survival Trust

In Britain the Rare Breeds Survival Trust was established in 1973 to prevent further extinction of British farm animals such as sheep, pigs, horses, goats and poultry, and to supervise the nation's conservation effort. Its aims include:

Above: The Cotswold Farm Park was the first Rare Breeds Survival Trust Centre in Britain.

- To build and maintain an information bank.
- To preserve selected breeds by encouraging and helping the owners of such animals.
- To support research to ensure that the gene bank created is used for the improvement of breeds currently in use and for the development of new breeds to meet future needs.
- To educate and conserve for the benefit of a wider public. Rare breeds are just as important as ancient buildings.

The Trust's work, which is supported entirely by subscriptions and donations, involves:

- Blood typing to identify relationships within breeds;
- Provision of registration facilities and herd and flock books to record all relationships;
- Formulation of breeding programmes based on knowledge of bloodlines within breeds;
- Collection, storage and distribution of rare-breed semen;
- Study prior to their utilisation of new scientific techniques which may assist genetic conservation;
- Arranging for individuals to keep groups of rare-breed animals on farms in suitable areas when an undesirable degree of geographical concentration exists;
- Providing means of communication between breeders;
- Organising and supporting a national show and sale to aid communication and facilitate the exchange of bloodlines between breeders at a national level;
- Publicising the cause of rare-breed conservation;
- Raising funds for all Trust activities;
- Providing financial incentives to individuals to keep rare bloodlines.

Membership of the Trust is available to the public, and members receive a monthly newsletter. The Trust also publishes a list of approved farm parks that display rare breeds. At the time of writing there were 15 such parks spread throughout England, and two in Scotland.

This information was provided by the Rare Breeds Survival Trust. To find out more about the Trust's work, see www.rbst.org.uk

The Rare Breeds Conservation Society of New Zealand

As animal breeders have constantly sought to improve their livestock, some of the original breeds have dwindled to low numbers and even died out. In Britain alone over 20 breeds of farm animals have become extinct since 1900. Worldwide, the rate of loss has been estimated as something like one breed per week, so the genetic diversity of livestock species is rapidly becoming reduced. A common breed can become rare and then extinct in a very short time unless someone is caring for it. This has happened to sheep, cattle, goats, pigs, poultry and horses.

Above: Rare Breeds Society Logo.

From the very first European landings by Jean François-Marie de Surville and James Cook in the late 1770s a steady stream of livestock has been introduced into New Zealand. Most of it came from Australia and Great Britain, but more recently importations have been made from countries all around the world. Many of the early breeds have been superseded by 'improved' varieties that better suit changing consumer demands and more efficient methods of farming.

During the early days many animals escaped into the wild and survived as independent breeding groups. During the 19th and early 20th centuries, cattle, goats, pigs and even rabbits were deliberately released on a number of New Zealand's offshore islands to provide food for shipwrecked sailors as well as stranded whalers and sealers.

In recent years efforts have been made to wipe out this feral livestock, both on the mainland and on islands, largely because of the damage they cause to native vegetation. However, it is recognised that many of these feral animals, especially those groups that were established during the 19th century, have had little or no contact with modern livestock breeds, and may have important genetic characteristics that have been lost in today's domestic breeds. Representatives of these feral groups have been rescued and are now being farmed throughout the country.

Enderby Island cattle are probably the world's rarest cattle breed. They are the descendants of 'shorthorn' cattle introduced onto subantarctic Enderby Island over a hundred years ago. In 1993 only one cow, 'Lady', survived from a herd of almost 50 after the rest of the herd had been culled for conservation reasons. The cow on the left of the photograph is a clone that was born in 1998, and the other two are daughters of clones from natural service, born in 2002.

For further information refer www.rarebreeds.co.nz/enderby.html

With the advent of DNA typing it is becoming possible to establish the relationship of feral livestock with existing domestic breeds. However, at this time the priority is to preserve the existing genotypes of these breeds and ensure their conservation – not primarily as relics of the past, but as something valuable and unique to New Zealand.

There are a variety of reasons for preserving these original and feral breeds:

- The original breeds still represent unique 'genetic packages' that may be called upon again some time in the future.
- Market fashions change. Lean sheep are now favoured over fat, and goats are now valuable to many farmers who used to shoot them as pests.
- Rare breeds often retain desirable characteristics that can be incorporated in new improved breeds of livestock.
- Genetic engineering is rapidly revolutionising breeding work but it still needs the appropriate genes on which to draw.
- Rare and unusual breeds provide material for research on the evolution of domestic characteristics.
- Feral types illustrate how animals change when they run wild, and represent aspects of New Zealand's colonising history.
- With much of the world being afraid of climate change, there is all the more need to conserve breeds and resources for an unknown future.

The Rare Breeds Conservation Society of New Zealand was formed in 1988 to facilitate the conservation, recording and promotion of these rare and minority breeds for the purpose of maintaining genetic diversity within New Zealand's livestock species. It has established a Rare Breeds Gene Bank to help retain valuable genetic material by cryopreservation.

The aims of the Rare Breeds Conservation Society are:

- To catalogue the holdings and locations of rare and unusual livestock.
- To foster official and amateur efforts to preserve rare breeds, special breeding groups, and feral types.
- To publicise livestock genetic conservation.
- To maintain contact with overseas societies and international agencies.
- To organise rescue and breeding programmes.

Many people find unusual animals attractive to look at, to breed, to keep or to exchange with friends, but on their own they can do only so much. By belonging to the Society they can develop contacts, share experience, get advice, build up their knowledge and avoid mistakes and losses. As a group, lay people, scientists and managers can do much more for the preservation of rare animal genetics than individuals working alone. And they can develop a common voice in the interests of endangered livestock.

This information was provided by the Rare Breeds Conservation Society of New Zealand. To find out more about the Society's work, see www.rarebreeds.co.nz

Genetics on the farm

Anyone who wishes to become seriously involved with cattle, whether breeding, commercial farming or merely as a hobby, needs a good knowledge of genetics. The subject receives only a brief mention here since it is not within the scope of this book, but some examples may assist the reader to understand better why breeds look and perform the way they do.

Two particular farming programmes that make use of genetics are grading up and genetic recovery.

Grading up

A grading-up programme uses foundation females of any type of breed or cross. By breeding successive generations to a pure-bred bull of a selected breed, stock can eventually be obtained that will be classified as 'pure-bred'.

Some breeds require a minimum of four generations of such breeding (15/16 purity) before registration; others require five (31/32).

Left: Mandalong Special heifers, part of a grading-up programme.

Genetic recovery

This is a scheme used by many breed societies worldwide. It recognises the value of genes existing among well-recorded but non-pedigree cows that have been sired by a pedigree bull, and aims to bring those genes into the breed's genetic pool. A registered sire is mated with those cows, and later their offspring up to the fifth generation, at which stage that animal is registered in the herd book.

Genetics in action

The genes that exert an obvious effect include those determining coat colour. When two different coat-colour genes are combined, one may exert its effect and mask the presence of the other. The masking gene is dominant, the masked gene recessive. For example, when black Angus are crossed with Hereford cattle, the offspring are black, but have the Hereford white face.

Dairy and beef production
Systems of farming

A variety of cattle farming systems are employed, and for descriptive purposes they are classified here to reflect the degree of husbandry involved. These classifications may not match those used in the field of agriculture, and many farms may employ a mixture of such systems.

In **primitive forms** of farming, cattle forage freely over unimproved and unfenced land with a minimum of herding.

Extensive systems operate on many cattle stations in the warmer parts of New Zealand: beef cattle forage widely over large areas, ranging from a few hundred to many thousands of hectares, and are never housed.

Semi-extensive systems allow cattle to forage widely over upland or mountain areas during summer months but confine them to more sheltered pastures or yards during the winter months. Areas where these types of system are found include high-country farms bordering the Southern Alps of New Zealand.

In **semi-intensive systems** some beef cattle and most dairy cattle are rotationally grazed on high-quality pasture but are also given supplementary

Below: Many of the cattle in New Zealand are never housed.

Above: Angus cattle being mustered on Tui Station.

food such as hay or silage. Milking cows may also be fed concentrates according to their productivity and the season of the year. In many areas of New Zealand, cattle remain outdoors all year round, but in colder climates they may be sheltered or even fully housed during the winter months.

Intensive systems keep cattle in closely confined areas such as barns, yards or feedlots, where they are fed high-quality foods such as grain. These systems are mostly used for beef production, and cattle may spend all their life in such areas, or be brought into them at a certain age for finishing.

Types of farmer

Stud breeders concentrate on producing pure-bred stock whose progeny are available for sale to commercial herds. For the stud breeder, success in the show ring can be important in creating or maintaining a good reputation.

Commercial farmers are the main producers of meat and milk. Some may also act as stud breeders, but the majority concentrate on breeding for production within their own herd.

Hobby farmers range from retired farmers who cannot quite let go, through to rural-dwelling company executives (and their husbands or wives), or families who simply want to live in the country and keep a few animals. Many of these people show an interest in exotic or unusual species and breeds, and on many of these farms may be found some of the less common cattle breeds.

Above: Jersey cows receiving hay supplementation during the winter.

Below: Winter housing in Britain – Devon cattle in a barn.

Types of cattle

Any tour of the countryside will soon reveal the diversity of cattle breeds and cross-breds that make up the rural cattle population. During the last decade there has been a significant change in the relative proportions of dairy and beef cattle, with dairy cattle (principally Holstein-Friesian) now in the majority.

New Zealand Sheep and Cattle Numbers
1990–91 to 2007–08

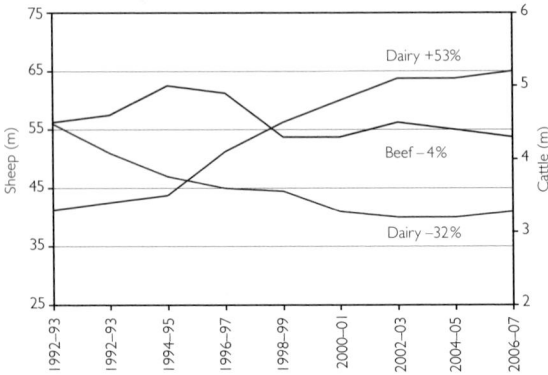

Source: Meat & Wool New Zealand Economic Service

This change to dairying is due to record prices for milk powder, caused by strong income growth in the Russian Federation and in many developing countries, especially in Asia, and by demand from oil-exporting countries.

However, the essence of the price spike lies on the supply side: global milk production has not kept pace with strong demand. Successive droughts in Australia have limited its milk product exports, while export taxes by Argentina have hindered its supply response. A six-month suspension of skim milk powder exports in India has eliminated its presence on world markets. High feed-grain prices have also curtailed profitability in many feed-intensive dairy sectors.

Finally, perhaps the most significant factor in recent times is current policy reforms in the European Union, which have resulted in a drastic reduction in public stocks of dairy products, especially milk powders, and a steep cut in the European Union's export subsidies, in both value and quantity terms.

Dairy farming

Dairy herds are not always composed entirely of one breed. For example, a particular herd might be largely Holstein-Friesian, but also have a few Jersey or Ayrshire cows. Although the cows in the herd have been selected for their milk production, the income generated by cross-bred calves is still an important part of the dairy farmer's business, so those cows not required to produce pure-bred replacement calves are mated or artificially inseminated using a beef breed bull. The sire of such a calf (destined for the market) is known as a terminal sire.

The resulting cross-bred calves are usually reared to a certain stage, often about 9 months of age, either by being weaned off their dam and fed milk or milk substitute until fully weaned, or by being fostered onto a suckler cow. A suckler cow might be one of the pure-bred dairy herd, but could well be a cross-bred, one of the most popular crosses being the Hereford × Friesian. The suckler cow is specifically selected for suckling and rearing one calf (single-suckling) or maybe more (multiple-suckling), the requirements for this being a docile temperament, very good mothering ability, and, of course,

Below: Dairy herds are not always composed of one breed.

a plentiful supply of milk with high milkfat and protein content. If the calves are to be weaned early and fed on concentrates, such a cow may rear as many as four calves.

If the farmer is considering a change to the breed profile of the herd, or experimenting with embryo transplant (say for the production of pure-bred Maine-Anjou calves), the same type of cow will probably be used as the recipient for the fertilised Maine-Anjou ovum.

Left: Three examples of pure-bred calves born to recipient Friesian × Hereford cows.

(*top*) Friesian × Hereford cow with Charolais calf.

(*middle*) Friesian × Hereford cow with Mandalong Special calf.

(*bottom*) Friesian × Hereford cow with Texas Longhorn calf.

Beef farming

Just as most commercial dairy farms have more than one breed, so too do most commercial beef farms, and it is common to see a mixture of pure-bred and cross-bred animals being reared together in a beef herd. In recent years there has been very strong interest in the use of cross-breeding programmes to introduce hybrid vigour into beef offspring and improve their growth performance. The options for such crosses are many, four of the more common being Hereford × Aberdeen-Angus, Hereford × Holstein-Friesian, Simmental × Friesian and Simmental × Hereford.

There are a number of different cross-breeding systems, including two-breed rotation, three-breed rotation and the terminal cross. The latter involves mating a cross-bred beef cow (e.g. a Hereford cross) to a heavily muscled bull of a third breed, a system which can give the maximum advantage of hybrid vigour. However, the offspring of a terminal cross system are usually sent to market, so this system is not self-sustaining. A recent development is to keep the heifer calves of such a cross, mate them to a terminal sire, and then send them off to market. In France, for example, the lower-graded Charolais are mated to a terminal sire, and their heifer progeny kept until they too have calved, after which they are sold at about 3 years of age.

The calves resulting from these systems are usually reared by their own dam (single-suckled), but a very high-producing cow may rear two (multiple-suckled).

Left: A Friesian × Hereford cow with her cross-bred Maine-Anjou calf.

Above: Aberdeen-Angus cattle in the Rangitata valley.

PART TWO: Cattle anatomy and behaviour

Cattle senses

The three main methods of communication for cattle are through sight, sound and scent.

Vision

Cattle are a prey species, so their eyes are set on the sides of their head to give them wide peripheral vision, which is in excess of 300 degrees. This can be further enhanced by movement of the head, neck or body.

A good side view is useful for detecting a threat, and when grazing with head down it is also useful for keeping an eye on other animals in the herd. When looking straight forward cattle do have a narrow 'blind spot', so to help overcome this they keep moving around. To inspect something, or examine a threat, they use their binocular vision which is about 25–50 degrees wide. Their eyes are designed to see down rather than up, so when alarmed they raise their head to investigate.

Although cattle have depth perception, they have difficulty perceiving depth at ground level while they are moving with their heads raised. To see depth at the ground they have to lower their head. They tend to baulk at moving or flapping objects, and also at shadows or distractions at ground level, lowering their head to investigate. They generally avoid bright light if given preference.

Cattle have colour vision, but it is not as good as ours. They can recognise different people from their shape and colour of clothing. They can also count, and can associate more than one person, or someone in green overalls, with the stress of forced handling or the pain of injections.

Far left: A good side view is useful for detecting a threat.

Left: Cattle can recognise different people from their shape and the colour of their clothing.

Hearing

As a prey species, cattle need good hearing to keep safe, and use their ears to amplify and pinpoint sound. Sound arrives at each ear at slightly different times and is processed into a directional signal, which can be further refined by moving the ears, head or entire body.

Cattle are sensitive to and aroused by high-frequency sounds which people cannot hear. Low-frequency sounds are more relaxing for them. In milking parlours music is regularly used to provide cows with a familiar background noise.

Cattle make a number of distinct sounds (vocalisations). Amongst these are the roar of an angry bull, the call of a cow to her calf, and the 'baa-ock' play call of a calf.

Smell

Cattle have a better sense of smell than humans, and use it in a number of different ways.

A cow leaves her young calf lying down in a safe spot while she goes off to graze, and uses her sense of smell to find it again. A bull uses smell to detect cows in heat (oestrus). The classic head-raised, lip-curling behaviour of the bull as he smells oestrus females, the Flehmen response, is generally exhibited after smelling freshly excreted urine. Its primary purpose is to expose the vomeronasal organ, located in the nasal cavity and connected to the roof of the mouth. This detects the pheromones (sexual odour compounds) in the urine. In turn, female sexual cycles are affected by the pheromones produced

Below: (*left*) Murray Grey cow calling. (*right*) Brahman bull displaying Flehmen response.

by bulls. The odour of urine also enables cattle to identify others of their species.

Some pheromones are released as a result of stress or fear, and can result in behavioural changes such as anxiousness, protective instinct or aggression.

The smell of blood can sometimes cause great panic; for example, when cattle pass paddocks treated with blood and bone fertiliser.

Taste

The taste buds of cattle are located mostly at the back of the tongue. When presented with a variety of feeds cattle will select certain feeds over others, and taste may play a role in this behaviour.

Touch

Cattle have a very sensitive skin, and touch is an important form of communication. Cows lick and groom their calves right up to weaning, and individuals participate in mutual grooming, which is especially prevalent in mature animals. Muscles under the skin enable cattle to flick flies off from localised areas.

It is important for handlers to use touch to warn cattle where they are; for example, when cows are being milked. One really bad experience will put cattle off all people for a considerable time until a positive human–animal bond is restored.

Below: Guernsey heifers mutual grooming.

Social and territorial behaviour

Herd structure

Wild cattle species today generally live in herds of 5–20 animals, each herd led by a bull and consisting of its females and their young. At about 2 years of age young bulls are banished from the herd and join a male sub-group. They fight for dominance, and eventually may be able to mate with some cows or even become 'king bull' and establish their own herd. Old males that have been deposed live alone.

Domesticated cattle herds are influenced by human management and the 'wild' social patterns have been largely or totally eliminated. They are still visible in the herd of wild cattle at Chillingham, in Northumberland, which have been subjected to a minimum of management.

Some aspects can be witnessed in free-ranging herds, particularly those running on the large beef-cattle stations in Australia.

Cattle follow the leader, and are motivated to maintain visual contact with each other. An animal separated from the herd will try to rejoin it.

The flight zone

This is the 'personal space' around cattle; if it is entered by a perceived threat, the cattle will react. Given a choice, they will move or run away before stopping, forming a group and facing the threat. If the threat approaches closer they will scatter and regroup again. They may naturally circle, perhaps to maintain visual contact with each other.

Social ranking – dominance and submission

Cattle herds have distinct hierarchies. Factors involved in determining these include sex, age, size (weight rather than height), the presence or absence

Right: Chillingham cattle.

Facing page: (*left*) Jersey bulls sparring. (*right*) A Dairy Shorthorn bull 'dusting'. Dairy bulls tend to be more aggressive than beef bulls.

of horns, and territoriality. In an established herd the older cows tend to dominate the younger ones, and the heavier cows tend to dominate the lighter ones. A horned cow will dominate a polled one.

Aggression is expressed by bunting or striking an opponent with the head. Aggression does not normally increase as the space available for each animal decreases.

Once a hierarchy has been established it usually remains stable and aggression is thereby reduced. The introduction of strange (in particular adult) animals can result in renewed aggression and associated stress until a new hierarchy has been established, after about 48 hours.

Groups of calves also develop a social order, apparently dictated by age and size. In dairy cattle this order lasts until the young animals are integrated with the main herd.

There are also breed differences in dominance: for example, Angus and Brahman cattle are usually dominant over Herefords, and Ayrshires are usually dominant over Holstein-Friesians.

Territory and aggression

Bulls become more territorial with age. Young bulls under 2½ years old tend to live together amicably, but by 3–4 years of age a clear dominance order has been established through trials of strength that include mounting and combat. In certain areas sparring holes are dug for threat behaviour and display, the bulls thrusting their heads into them to cover their heads and horns with dust or mud depending on the weather conditions. In dry conditions dust is also thrown high into the air and over the back and withers by pawing the ground with the forefeet. At the same time the bull will vocalise.

Because of their innate impulse to become dominant, domesticated bulls

Above: (*left*) A bull's roar: an Australian Shorthorn announces his presence.
(*right*) Red Poll cow chewing cud.

are notorious for their unpredictable aggressive behaviour, although those running with cows or groups of other bulls pose fewer problems than those that are housed alone, receive minimum exercise and are unable to see what is happening around them. Dairy bulls tend to be more aggressive than their beef counterparts. Cows can also be aggressive when protecting their calves, especially the newborn.

Daily routine
The pattern of the normal bovine day depends on the diet or pasture conditions, and the type of housing.

Wild cattle generally rest early in the morning and in the middle of the afternoon, when they chew the cud (ruminate). In between these times they graze. They normally sleep at night, but in areas where they are disturbed they may adopt a nocturnal lifestyle.

Domestic cattle are essentially active during the day (diurnal). They spend 5–8 hours of the day grazing, and about the same amount of time ruminating. When grazing, the animal's head moves through an arc of 60–80 degrees. Grazing usually involves two major bouts, one from sunrise to mid-morning and the other from late afternoon until just after sunset, with a third, shorter period around midnight. In lactating dairy cattle these patterns are usually disrupted by the milking schedule, and animals may need to graze 1–4 hours during the night to compensate. On moderately good pasture non-lactating cattle spend about 5 hours grazing and 2 hours walking. The distance covered can vary from just over half a kilometre to 9 kilometres, depending on the size of the area and the abundance of forage. Cattle will often make and use clearly defined pathways or tracks.

Lactating dairy cattle normally graze for 8–10 hours daily.

Rumination occurs mostly while the animal is lying down, and there are normally 15–20 periods of 'cud-chewing' during a 24-hour period. Rumination time decreases under hot conditions.

Compared with cattle on pasture, those in loose housing spend only about half as much time eating and ruminating (5 hours in total), a further 7 hours loafing (neither eating nor ruminating) and 12 hours resting. Cattle in feedlots are in a highly unnatural environment and feed 9–14 times during a 24-hour period, mainly during the day, the amount of time spent eating decreasing as the proportion of concentrates increases.

In many breeds of cattle, food intake increases during cold weather and decreases under hot conditions such as those experienced in the tropics. Cattle breeds adapted to the tropics (zebu, zebu-cross and Sanga) are better able to graze in the heat, but may also eat more at night when the temperatures are cooler.

Above: Cattle walking along a track.   *Below*: (*left*) Devon cattle resting. (*right*) With food scarce, this Brahman heifer is grazing along railway tracks.

Above: (*left*) Holstein-Friesian cow drinking. (*right*) Longhorn calf sleeping.

Drinking

During summer cattle usually drink two to four times a day, but in winter this decreases to once a day or even less. A high-producing dairy cow requires about 160 litres of water daily.

Sleep

Within an hour of sunset most cattle are lying down. They spend about 4 hours sleeping, taken in a series of short naps. REM (rapid-eye-movement) sleep accounts for about 45 minutes of this, usually characterised by the animal lying down with its head resting on the ground, turned back into the flank. Cattle may get up to graze at some stage during the night, particularly during warm weather.

Sun and shade

In hot climates or conditions cattle will move into shade if it is available, and for optimum comfort all cattle should have access to it. Cattle probably seek shade in response to light intensity rather than the ambient temperature. Tolerance to heat and strong sunlight varies according to breed and colour of skin pigmentation. The tropically adapted humped breeds have good heat tolerance and a high proportion of sweat glands in their dewlap and hump. Breeds with dark skin pigmentation, such as the British White, are less prone to sunburn and eye cancer.

Defecation and urination

Cattle defecate 7–15 times per day and urinate 5–13 times. The frequency of both activities decreases during hot weather.

Maintenance and grooming

About 1 hour daily is spent in licking. Some mutual grooming occurs but only occupies a few minutes of each day. Older and larger animals give and receive more grooming than younger animals. Subordinate animals lick dominant ones around the head and neck.

Breeding behaviour

Puberty. Females reach puberty at 6–12 months of age. Cows are non-seasonal, continuously cycling breeders, but peak fertility occurs in spring and is low during mid-winter. The oestrus cycle is 18–23 days long, with a mean of 21 days, and length of oestrus varies from 6 to 23 hours, depending

Left: Hereford cattle in the shade near Kaikoura.

Right: Guernsey heifer grooming.

Above: (*left*) A Poll Hereford bull adopting the T-position with a cow in oestrus.
(*right*) Jersey cow with newborn calf.

on the season, presence and sex of other cattle, and feeding level. Males reach puberty at 5–12 months of age and are capable of serving a female at that age, although if they are in a herd with a dominant bull, the latter will prevent them from doing so.

Mating behaviour. As a cow approaches oestrus the dominant bull will start to graze beside her and guard her, using the Flehmen response to detect her pheromones (see Smell, page 54). During oestrus he will frequently maintain contact with her. This may be head-to-head, head-to-tail, or standing in a T-position with his head across the cow's back.

Pregnancy and birth

Gestation averages 282 days, but is highly variable between breeds. It is longer in some European breeds; for example, in the Limousin it is 5–10 days longer. Gestation is not just a function of the female; the sire also has an effect and may be deliberately selected to reduce its length.

Approaching birth, a cow will move to a quiet sheltered place at the edge of the herd. After birth, bonding between cow and calf occurs quickly, often within 5 minutes. The wet newborn calf is licked by its mother, who cleans off and eats much of the foetal membrane and amniotic fluid that remains on the calf.

If contact between a cow and her calf is delayed for 5 hours, in 50 per cent of cases the calf is rejected. Cows groom their calves during the early post-partum period, concentrating on the back and abdomen.

Above: (*left*) Simmental calf suckling. (*right*) Jersey calves in a crèche.

Nursing and suckling

Most calves can stand within an hour, and nurse within 2 hours. It may take 20–30 minutes for the calf to locate the teats. Calves adopt a particular stance while nursing, and butt upwards at the udder to stimulate milk flow. They also wag their tails. Newborn calves usually nurse five to seven times a day, mainly during daylight, each session lasting about 10 minutes. In dairy breeds the frequency of suckling tends to decrease as the calf gets older, whereas in beef breeds it may actually increase.

Cow and calf daily rhythm

Cows and calves spend the night lying together in groups. When the cows get up the calves suckle, then leave their dams and lie down together forming a crèche while the cows graze. About mid-morning one or two calves will start to bawl, and this triggers a mass return of cows to suckle their calves. The same pattern occurs during the late afternoon with the cows returning to suckle their calves prior to nightfall, by which time they have reformed to lie down as a group.

Play and investigation

Calves (and indeed older cattle) are naturally inquisitive. They enjoy periods of play, which can average about 3 minutes a day at 7 weeks of age. Play is often stimulated by a change in the environment, such as introduction to a new paddock or the placement of new bedding in a stall. Calf play includes trotting or galloping with the tail elevated, head butting, kicking sideways and upwards with both hind legs, prancing, pawing at the ground, and mounting and pushing. Male calves mount, push and butt more than females. Calves will also play by making a noise with an object such as a bucket. The amount of time spent in play decreases as the calf gets older.

Handling and moving cattle

The flight zone is an animal's personal space: if you move inside it the animal will move away. If you back off, it will stop moving. The size of the flight zone depends on the angle of your approach, the wildness or tameness of the cattle, and their state of excitement. The flight zone radius may be 2–8 metres for feedlot cattle and 100 metres for range cattle. Work at the edge of the flight zone at an angle of 45–60 degrees behind an animal's shoulder – called the point of balance. If you are behind the point of balance, the animal will retreat and circle away from you.

A technique that works well with cattle that are moving is not to approach them directly, but to work close to the point of balance, moving back and forth on a line parallel to the direction the animal is travelling.

Some key points for moving cattle are:

- Handle small groups in crowding pens: 8–10 animals rather than 20. The cattle need room to turn.
- Cattle are motivated to maintain visual contact with each other. Exploit this by making sure that each animal in a chute can see others ahead of it.
- Blocking gates in a chute need to be 'see-through'. If a cow sees a dead-end it will baulk, and will then continue baulking.
- An animal left alone in a crowding pen will become agitated and may attempt to jump the fence to rejoin its herd-mates. If a lone animal refuses to move, release it and bring it back with another group.
- Cattle will baulk at striped shadows and contrasting patterns, so use uniform lighting in handling areas.
- Cattle in the dark will move towards the light.
- Cattle may refuse to enter a dark, indoor working chute from a bright, outside crowding pen.
- Cattle tend to baulk if they have to look into the sun, so orient your cattle chutes accordingly.

Above: Ayrshire calf playing.

Aspects of cattle anatomy

Beef animals are bred for heavy musculature, or for double-muscling, as exemplified by this Belgian Blue. The double-muscling is evident in the ripples beneath the skin.

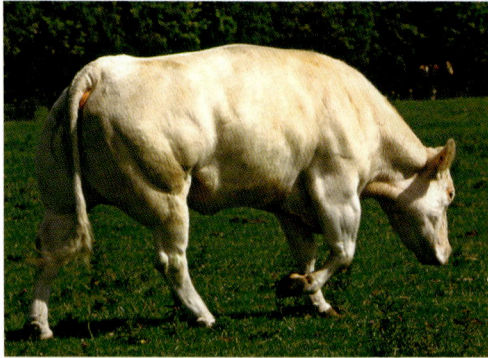

In contrast to beef breeds, dairy cattle are bred to put all production into milk. Docile and placid, Jersey cattle exemplify the classic dairy breeds. Although ribs and hip bones can be seen through the skin, this animal is not malnourished: the dairy cattle breeds simply have no need for heavy muscle cover.

Note the prominent mammary vein, teat placement, and distension caused by copious milk production on the udder of this Holstein-Friesian. These factors have been accentuated by generations of careful selective breeding.

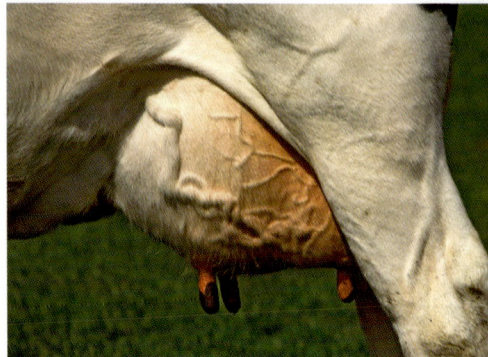

The calves of some breeds, such as the Marchigiana, change noticeably in colour as they grow. By the time it is 6 months old, this calf's coat will be the same colour as its mother's.

Sahiwal cattle have been bred for resistance to relatively hot, dry climates. Descended from *Bos t. indicus* rather than *Bos t. taurus* ancestors, their body shape is very different from that of the traditional European dairy and beef breeds. Distinctive features are the dewlap – the loose fold of skin running from the neck down to the chest, which aids cooling – and the hump (the function of which is not known, although it is composed mainly of muscle tissue and contains a high proportion of sweat glands). *Bos t. indicus* bulls tend to have more prominent pizzles than *Bos t. taurus* bulls; many cows also have what appears to be a pizzle but is in fact a large umbilicus.

Some breeds moult seasonally – in this case, Galloway cattle. In winter their coats are very long and shaggy, almost like a sheep's wool rather than an ordinary cattle coat. The Galloway's summer coat is also fairly rough, evidence of the breed's northern British heritage.

A typical cattle hoof. Hooves also vary between breeds – animals that spend much of their time in swampy conditions will tend to have splayed, wider hooves to spread their weight more evenly across the ground; while animals in hot, dry climates, especially those which have to walk great distances foraging, have narrower, much harder hooves. The back toe serves no real purpose.

Breeding of horned and polled varieties has created the most readily distinguishable forms of cattle skulls. Even if an animal has been dehorned, the skull will still show the horned conformation. Note the wide spacing of the eye sockets: like other herbivores cattle have the ability to see in a wide circle around them.

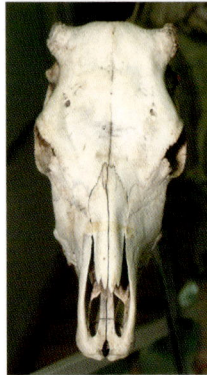

Horns come in a variety of shapes and sizes. The horn is made of keratin, like human fingernails only much harder.

The distinctive hooded oblong eyes of tropical breeds developed as protection against strong sunlight. Note also the long, droopy ears.

In contrast, this European animal's ears are much more upright, and the eyes are rounder and less hooded.

Beads of dew can be seen on this Holstein-Friesian's nose.

Cattle have long, very muscular tongues, which they use to grasp and pull grass with when grazing – and also in grooming.

Cattle have eight lower incisor teeth. In place of the upper incisors there is a hard dental pad.

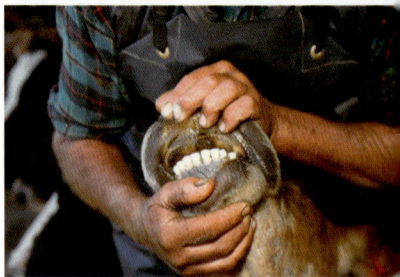

Like many animals, cattle have whiskers around the nose and mouth which act as touch sensors.

PART THREE: Cattle breeds

Key	
(D)	Dairy
(B)	Beef

Angus (Aberdeen-Angus) .... (B)

Developed in the early 19th century from various strains of hardy polled cattle in the counties of Aberdeen and Angus in north-east Scotland, the breed was formally recognised in 1835. It rapidly expanded during the industrial revolution, when steam shipping and trains enabled Scottish breeders to send their cattle to new markets in England. The north-east of Scotland became the main beef-producing area in Britain, a position which it still holds. During the late 19th century the breed spread rapidly throughout Britain and Ireland and to all the major beef-producing countries of the world.

First imported into New Zealand in 1863, the breed has been transformed from its smaller dumpy conformation to a moderately large, well-muscled and leaner breed, known simply as the Angus.

Within New Zealand the pure-bred Angus is numerically the largest beef breed, enhanced considerably by Hereford and Shorthorn cross animals. At the end of 2007 registered Angus females numbered 23,000, including 2-year-old heifers, almost all of them calving down as 2-year-olds.

A small version of the Aberdeen-Angus, the Australian Lowline, has now been developed (see page 78).

APPEARANCE

A moderately large and full-bodied breed, with good muscling. Naturally polled. Coat black. Some animals carry one red gene, masked by the dominant black colour. In certain light or when animals are in winter coat a reddish tinge may be visible.

Left: Angus bull.
Right: Angus cow and calf.

BREED FEATURES

Very hardy, a good forager, and able to thrive on low-quality pasture or cheap rations such as silage or arable by-products. Particularly suited to extensive grazing. Females are relatively long-lived and make ideal suckler dams in upland or lowland conditions. A very popular and successful beef breed.

Breeding

High fertility and regular breeding. Cows hardy and thrifty, easy calving due to moderate birth weights. Calf birth weight averages 43 kg. Cows display good mothering ability. Bulls are promoted as excellent terminal sires for cross-breeding with dairy as well as other beef breeds.

Growth and carcass

Superior carcass quality with a high proportion of lean meat to fat and bone, and very well developed hindquarters producing a high proportion of the most expensive cuts, such as sirloin and rump. Meat lean and well marbled, with moderate fat cover which imparts flavour and juiciness when cooked.

Height and weight

Mature bull:	Average height at withers 137 cm	Average weight 950 kg
Mature cow:	Average height at withers 130 cm	Average weight 750 kg

Red Angus

(B)

Derived from the black Aberdeen-Angus breed, in which many animals carry a red recessive gene that is masked by the dominant black gene. Red-coloured Angus were recorded as long ago as the early 19th century, and in many countries, including Britain, they are accepted and registered in the same breed register as black-coloured Angus. However, in the United States red animals were barred from registration after 1917. Various cattle farmers throughout the United States saw merit in the red animals, and set up breeding programmes. By 1954 a sufficient number of herds had been established to form a breeders' association known as the Red Angus Association of America, which was the first beef breed organisation in America to make performance recording mandatory and a requirement of membership. Since then red animals have competed successfully against, and sometimes surpassed, their black counterparts.

BREED FEATURES

All the genetic attributes of the Angus. It is claimed that the red colour results in better heat tolerance, and the bronze pigmentation improves resistance to eye cancer and sunburn on the udder.

Breeding

The red factor also enables commercial cattle breeders to cross Red Angus with other red-coated breeds (e.g. Hereford, Simmental, Limousin, Salers and Gelbvieh) and maintain the red coat colour. In American and Australian commercial herds there has been considerable cross-breeding of Red Angus with the Hereford (producing the 'red baldy') and the Simmental (producing a blaze-faced animal). Similar breeding programmes are now taking place in New Zealand.

APPEARANCE

Size and conformation similar to the Angus. In America breeders have concentrated on producing animals of moderate size and eliminated those that are comparatively small. Coat colour red. Naturally polled.

Left: Red Angus cow and calf. *Below*: Red Angus bull.

Aubrac Ⓑ

The breed is named after the most southern of the volcanic masses of Auvergne, on the southern slopes of the Massif Central in south-central France, where the terrain ranges in height from 500 to 1500 metres above sea level. The climate can be severe in winter, with snow sometimes lasting from October to May, and very hot in summer. A temperature difference of 30°C can be recorded within a few hours.

In the 17th century the Benedictine monks at the Abbey of Aubrac cleared away much of the large, thick forest covering the plateau, and it was they who organised the first rationally managed herd. In 1888 a special commission defined the breed 'standard', and in 1894 an association of breeders was formed whose members kept records of their cattle and their progeny. This is now the herd book of the Aubrac breed. Although milk production was recorded, the Aubrac was then still considered as a work breed. These days it is regarded as a meat producer, although a minority of herds still milk for cheese.

The first imports into England arrived in early 1990, the objective being to export semen and embryos to the United States. This achieved, bulls and semen are now being sold to British milking and commercial beef herds, with particular interest coming from Jersey breeders.

In New Zealand the first calves were born in September 1994 following the importation of four live animals from America's sole herd, which itself originated via the import of four live embryos from France, and from embryos sent to the United States from the British herd.

The Aubrac is likely to become more popular with time.

Right: Aubrac bull.
Facing page: Aubrac cow and calf.

APPEARANCE
A stocky, medium-sized breed. Colour light to dark fawn, with dark areas around the eyes, inside the ears, on the feet and tail tuft.

BREED FEATURES
Hardy and adaptable to a wide range of climatic conditions. The breed has not been selected to be competitive in intensive operations, but was developed for difficult conditions where fodder is a limiting factor. It is now attracting attention for cross-breeding.

Breeding
High fertility in both males and females. In their native region heifers are not calved until late winter (Jan–Feb), when they are 3 years of age and housed: they are then ready to go up the mountains for the summer months (May–Oct) with their calf at foot. Ongoing fertility is very good, almost 100 per cent of mated females weaning a calf, with a calving interval of 12 months. In England heifers produce their first calf at 30 months of age. Calves are born vigorous, average 31 kg in weight, and are excellent survivors. Cows are relatively long-lived, with a few still calving at 17 years of age although the mean is 11 years. Because it can thrive on low-grade natural grassland the Aubrac is gaining the interest of commercial beef producers in Australia, New Zealand, Britain, Ireland, Canada, Mexico and the United States, particularly for cross-breeding.

Growth and carcass
Aubrac cattle reared under extensive conditions on low-grade pasture in harsh environments show a live weight gain of about 550 g a day. This rises to about 1250 g daily when animals are reared on grass.

Height and weight
Mature bull:	Average height at withers 139 cm	Average weight 900 kg
Mature cow:	Average height at withers 129 cm	Average weight 750 kg

Australian Lowline .... Ⓑ

The newest indigenous cattle breed in the world, the Australian Lowline was developed at the Trangie Research Centre, New South Wales, Australia. An experiment was commenced in 1974 to evaluate the effect of selection for growth rate. The Centre's herd of Aberdeen-Angus cattle, closed since 1965, was divided into three closed lines: the High line, which selected for larger animals; the Low line, which selected for smaller animals; and the Control line from which individuals were randomly selected for both High and Low lines. By about 1989, after several generations, distinctly different types of animal had been developed in the High line and Low line herds. Selection continued until 1992 when the University changed policy and the Low line cattle were sold. The owners of these cattle then formed the Australian Lowline Cattle Association in 1992. Three years later there were nearly 130 members, and the Australian herd numbered close to 500 females and 100 bulls.

Lowlines entered New Zealand in 1995 and in October of that year there were three bulls and 22 cows. Today there are approximately 500 active registered pure-bred Lowlines in New Zealand, and a far greater number of pure-bred Lowlines run as commercial animals in beef herds.

APPEARANCE
At maturity Australian Lowlines are two-thirds the size of standard Angus cattle, from which they were derived. Naturally polled. Coat colour black.

Above: Australian Lowline cow and day-old calf. *Facing page*: Australian Lowline bull.

BREED FEATURES

A docile, hardy beef breed whose smaller size allows for higher stocking rates and smaller, low-cost handling facilities.

Breeding

Good fertility. Easy calving. Calf (pure-breds) birth weights range 18–28 kg. Good longevity.

Growth and carcass

Most calves double their birth weight by about 6 weeks of age. By 12 months heifers average 250 kg and bulls 280 kg; at 18 months heifers average 350 kg and bulls 400 kg. Carcass comparatively lean. Meat of high quality and well marbled. Cross-bred Lowlines are very popular for carcass competitions in Australia, winning many top Royal Show competitions with the quality of their meat.

Height and weight

Mature bull: Average height at hips 110–125 cm   Average weight 650 kg
Mature cow: Average height at hips 100–120 cm   Average weight 420 kg

Ayrshire .... (D)

Poorly built black cattle that were the ancestors of the Ayrshire roamed Scotland many centuries ago. During the 18th century breeders undertook improvements by interbreeding with West Highland, Shorthorn and Dutch cattle. The breed as we know it today came into existence in the area of Dunlop, in the Cunninghame district in the north of the shire of Ayr, in about 1794, although cattle of similar type are known to have been in that area for at least a century prior to that. The British Ayrshire Cattle Society was formed in 1877.

The Ayrshire was the second cattle breed to arrive in New Zealand, being first imported in 1848 when a bull and several cows were landed at Otago in the South Island. The first New Zealand herd book was published in 1910. Ayrshire New Zealand now has almost 400 members and there are approximately 20,000 registered pedigree animals in New Zealand. The association runs a 50/50 sharemilking venture known as the Ace Farm in Taranaki with the aim of improving the rate of genetic gain of the breed. Semayr Breeding Services, the marketing division of the association, markets proven and unproven Ayrshire bulls through LIC.

APPEARANCE
Medium-sized with naturally upward-curving horns. Most animals are now dehorned. Coat colour red, brown, mahogany and white. Brindle is not considered desirable but is not a disqualification. All animals have some white on the face or forehead.

BREED FEATURES

Hardy, a good forager, and capable of thriving during periods of stress in both hot and cold climates. Careful selection has resulted in a well-placed udder and long-wearing legs. Good temperament in and around the milking shed.

Breeding

High fertility, easy calving, good mothering ability. Calf birth weight about 40 kg. Calving index regular, allowing a calf every year and regular milk production. Good longevity, about 14 per cent of cows exceeding 10 years of age.

Milk production

Ayrshires are noted for their ability to produce very good yields of milk and milkfat under all conditions and over a long period (long-distance production). The milk has small fat globules and a natural soft curd, making it easily digested by humans and ideal for cheese-making. It contains high levels of protein, with a protein–fat ratio averaging 83 per cent, one of the best of the major dairy breeds.

Growth and carcass

Pure-bred calves grow quickly and have a good feed conversion rate. Bull calves develop into good dairy beef. Surplus breeding cows can be crossed with selected beef breeds to produce dairy beef offspring that have very good live weight gains.

Height and weight

Mature bull:	Average height at withers 150 cm	Average weight 950 kg
Mature cow:	Average height at withers 135 cm	Average weight 575 kg

Facing page: Ayrshire calf resting. *Above*: Ayrshire cow (*left*) and bull (*right*).

Belgian Blue or Belgian White-Blue .... Ⓑ

Accounting for about half of the cattle in Belgium, the breed was established in the early part of the 20th century, although for a century or more prior to that Belgian farmers had been crossing the native red-pied and black-pied cattle with Durham (Shorthorn) cattle (from which came the roaning gene) and the heavily muscled Charolais. Development of the breed was initially towards two separate strains, dairy and dairy/beef, but it is now primarily used for beef production. British and Danish animals do not have the extreme conformation of their Belgian counterparts, but their mobility and ability to cover the ground have been enhanced by selective breeding.

The Belgian Blue first entered Britain in 1982, and is now well established there. The principal use for bulls is as terminal sires for the production of cross-bred cattle for the vealer and domestic trade.

Belgian Blue semen was imported into New Zealand in 1984, followed in 1987 by the import of live animals and further semen. A breed society was formed in 1989. Bulls are used as terminal sires over all types of cattle, with the resulting cross-bred offspring producing a very high carcass yield.

APPEARANCE

A large but light-boned breed. Characteristic double muscling, a straight back, sloping rump, hidden hips and strong fine legs. Colour can be white, blue, blue roan or black and white. Naturally horned.

Right: Belgian Blue bull.
Far right: Belgian Blue calf suckling.

BREED FEATURES

Docile and easily managed. Growth rate and feed conversion are on a par with other beef breeds, but double muscling increases its production of meat.

Breeding

Gestation consistently normal at 281–282 days. Because of the comparative size of the calf, Belgian Blue cows are prone to calving problems. Although efforts are being made to select strains that have less difficulty during calving, in many herds some or all of the calves are intentionally delivered by Caesarean section. Calf birth weight averages about 45 kg, although weights of up to 65 kg have been recorded. The Belgian Blue can be used for cross-breeding with other beef breeds, and bulls are recommended as terminal sires for cross-breeding with selected large-framed high-yielding dairy cows, particularly Holstein-Friesians, thereby significantly enhancing the value of their offspring.

Growth and carcass

The double muscling ensures a very good carcass yield, with more than double the amount of prime cuts and very much lower fat and bone content than other breeds. The lean meat has cholesterol levels lower than those of fish or chicken.

Weight

Mature bull: Average weight 1200 kg
Mature cow: Average weight 800 kg

Belted Galloway ⓑ

A colour variant of the Galloway (see page 102) from the south-west of Scotland, developed from the ancient Celtic cattle. The white-belted animals were originally kept by the women of the clan for milking. The Belted Galloway was given the status of a separate breed when the herd book was established in 1921.

The Belted Galloway, together with the Galloway, first entered New Zealand in 1947. Both Belted Galloways and Galloways thrive best in cold, rough hilly country, so they are mainly limited to the colder southern tableland and mountain regions of New Zealand.

APPEARANCE
A medium-sized beef breed with a unique double hair coat comprising an outer layer of soft, shaggy, wavy hair and a soft mossy undercoat. Colour black or dun, with a distinctive belt of white.

Above: Belted Galloway cow and calf. *Right*: Dun Belted Galloway bull.

BREED FEATURES

Hardy and adaptable; can withstand harsh winter climates without the need for housing; comparatively disease-resistant; good conversion of roughage to meat. Said to be more docile than the Galloway.

Breeding
High calving percentage, easy calving. Calf birth weight about 33 kg. Cows are very good mothers, good milkers, regular breeders and long-lived.

Growth and carcass
A relatively small amount of subcutaneous fat results in a lean carcass with well-marbled meat.

Height and weight
Mature bull: Average height at withers 135 cm   Average weight 800 kg
Mature cow: Average height at withers 120 cm   Average weight 500 kg
North American bloodlines are taller and heavier.

Blonde d'Aquitaine .... Ⓑ

Cattle with light-coloured mucous membranes, probably descendants of *Bos aquitanus*, have existed in the south-western region of France for centuries. Two hundred years ago French cattle breeders recognised a large Aquitaine breed with five strains of equal importance – the Agenaise, Garonnaise, Limousine, Landaise and Urt – and early in the 19th century English breeders were importing stock from France to improve their own herds. The isolated nature of the French valleys resulted in strains evolving separately, with the formation of a Garonnais herd book in 1889, a Quercy herd book in 1920 and a Blonde des Pyrénées herd book in 1921.

Before World War II the breed was mainly used for draught, but the introduction of mechanisation resulted in a rapid decrease in numbers. In 1962 the three strains were merged to form the UPRA Blonde d'Aquitaine. The breed has now spread to all the principal livestock areas of France and numbers have increased significantly. Live animals and semen were imported into New Zealand in 1973, where the Blonde d'Aquitaine remains a minority breed.

APPEARANCE

A medium–large breed with well-developed body and bone structure, thick well-rounded muscle, broad back and exceptional body length. Coat colour wheat, in various shades from light to dark, often mottled, with lighter rings around the eyes and muzzle, inside the thighs, on the belly and over the lower legs. Mucous membranes are pale pink. Horns light-coloured, darker at the ends. Light-coloured hooves.

Right: Blonde d'Aquitaine cow and calf.
Far right: Blonde d'Aquitaine bull.

BREED FEATURES

Hardy, adaptable to a wide range of climates. Capable of foraging over long distances in rugged terrain, and converting poor-quality feed to flesh. Strong yet fine-boned. Very docile.

Breeding

Very good fertility. Easy calving due to a wide pelvis and the shape of the calf, which at birth is long and rather flat with light bones, an elongated head and small feet. Calf birth weight averages about 39 kg. Good maternal qualities and good longevity, with over 10 per cent of cows producing calves at more than 10 years of age and cows aged 15 years not uncommon. Bulls are particularly promoted for use as terminal sires in cross-breeding programmes with mature cows of other dairy and beef breeds.

Growth and carcass

Growth rates average about 1 kg per day under normal husbandry, but can reach 1.8–2 kg daily under intensive conditions. In growth trials the Blonde d'Aquitaine has performed particularly well. Carcass yield exceeds 60 per cent. Meat is very lean with a good proportion of prime cuts.

Height and weight

Mature bull:	Average height at withers 145 cm	Average weight 1050 kg
Mature cow:	Average height at withers 140 cm	Average weight 850 kg

Brahman .... Ⓑ

In the mid-1800s about 300 zebu (*Bos t. indicus*) draught animals were sent to the Gulf Coast of the United States from India and Brazil. These animals, which probably included the Ongole (also known as the Nellore), became the foundation stock from which the Brahman breed was developed. In the early 1900s some British-bred cattle were crossed with their progeny, and in 1924 a Brahman breed society was formed. The breed has become very popular in America, particularly in the warm southern States. Incidentally it is Brahman bulls that are commonly used in rodeos.

Worldwide the Brahman has proved extremely successful, and is one of the most numerous of all the cattle breeds. It has also been involved in the development of the Braford, Brangus, Charbray, Droughtmaster, Greyman, Mandalong Special, Santa Gertrudis and Simbrah breeds.

Brahmans were imported into New Zealand in 1994 but the breed has excited little interest, probably because of this country's equable climate.

APPEARANCE

A tall, long, medium-sized breed with pendulous ears, prominent dewlap, deep round body, strong sturdy legs, and well-muscled hindquarters. The large round thoracic hump is larger in the male than the female. The coat is sleek with short flat hair; the skin is dark-pigmented. The coat colour is predominantly grey, but red, brown and black colours also occur and are quite acceptable. Naturally horned, but polled strains are increasing in popularity.

Left: Brahman bull.
Right: Brahman cow and calf.

BREED FEATURES

Heat tolerant due to the highly developed sweating mechanism, while skin pigmentation protects against the sun. Resistant to internal parasites and ticks. Hardy, an excellent forager that will walk long distances for food and water. Highly intelligent and responds well to good handling and training. Temperament can be docile if handled properly but fractious if treated poorly or untrained. Cows with calves are highly protective of their young.

Breeding

Very easy calving due to the narrowness of the calf's body at birth. Calf birth weight about 40 kg. Excellent mothering ability with strong protective instinct. Very good milk production ensures good pre-weaning growth in the calf. The Brahman is widely used in various cross-breeding programmes.

Growth and carcass

Brahmans have a low maintenance requirement so they can grow well on poor-quality grasses and during periods of feed shortage because they are able to recycle essential nutrients through the bloodstream and saliva, which promotes digestion as well as a slower rate of protein turnover. The carcass has an even distribution of subcutaneous fat with less fat deposited between the muscles than British breeds. This increases the yield of saleable meat by 2–4 per cent.

Height and weight

The figures below are for well-grown cattle in good condition.

Mature bull: Average height at withers 162 cm   Average weight 950 kg
Mature cow: Average height at withers 142 cm   Average weight 600 kg

British White .... (B)

A native British breed whose origins are unknown although its polled status suggests some ancient Scandinavian influence. A herd of white polled cattle was known to be running in the 'Lords Park' at Whalley Abbey, Lancashire, at the time of the abbey's dissolution in 1537 and a herd was maintained there until it was dispersed in 1697. The breed spread to East Anglia which became the major centre from the late 1800s until the 1960s. A breed society was founded in 1918, when the first herd book was published. Originally the breed was of dairy type but was changed, particularly during the 20th century, by introgression from other breeds which accentuated the beef type. However, its popularity diminished and herds declined or were dispersed. By the mid-1970s only the Hevingham herd in Norfolk retained a considerable number of pure cattle, and overall breed numbers had declined to 100–150 pure-bred females. Old-established herds at Woodbastwick, Kelmarsh and Faygate also played an important part in the survival of the breed.

Today British White cattle are kept almost exclusively for beef production, and significant new herds have been established in most parts of the United Kingdom.

At the time of writing there were breeding-up programmes in New Zealand, using semen from the United Kingdom, which had produced some pure-bred animals.

Left: British White bull.
Right: British White heifer.
Far right: British White calf.

BREED FEATURES

Hardy and tolerant of a wide range of environmental temperatures and conditions. In regions of high ultraviolet light, the breed's pigmented skin and black points lessen the risk of sunburn, and the dark pigmentation around the eyes is effective in preventing eye cancer. Good milk production results in fast-growing calves.

Breeding

Good fertility. Easy calving, excellent mothering. Calf birth weight about 39 kg. Good longevity. Bulls are promoted for use as terminal sires on beef and dairy heifers.

Milk production

Now rarely used for this purpose, although official records show that during the 1930s yields averaged 875 gallons at 4.5 per cent milkfat.

Growth and carcass

A versatile breed for both extensive and intensive systems. Good weight gain and feed conversion. The meat is lean, fine-grained, and has a very good flavour.

Height and weight

Mature bull:	Average height at withers 140 cm	Average weight 835 kg
Mature cow:	Average height at withers 129 cm	Average weight 575 kg

APPEARANCE

A medium-sized breed with white (occasionally brindled) coat and black points on the ears, nose and lower legs. Red points occasionally occur. The skin colour tends towards dark grey. Naturally polled.

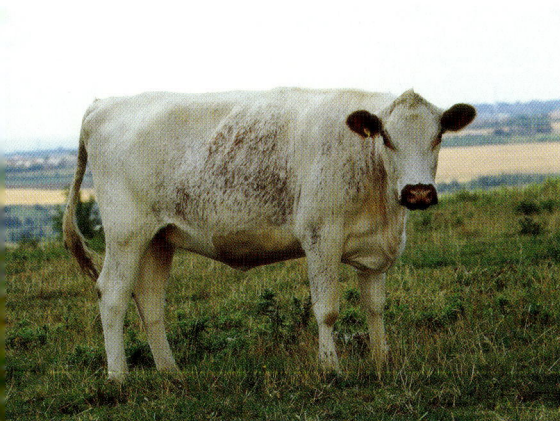

Brown Swiss or Braunvieh .... (D) (B)

Originating in Asia, the ancestors of these cattle were introduced to Europe during the 4th and 5th centuries. During the 17th century there were at least 12 different types of brown mountain cattle in Switzerland, but by the 19th century selection by the Swiss had made a more uniform breed. It is known locally as the Schwyzer, but generally called the Braunvich or Braunvieh in Europe, and by the English names of Brown Swiss, Swiss Brown or Brown Mountain. The breed was originally multi-purpose, used for milk, meat and draught. Selection for the last two purposes accounts for the muscling and docility that are features of the modern breed, while careful recording by Swiss monks of the cheese production from the milk of each cow incidentally selected for milk protein yield. The first herd book was started in 1878, and a breed society was formed in 1897.

In 1869 the breed was imported into the United States, and these strains have been selectively bred for dairy production since 1906. As a result the Brown Swiss and the European Braunvieh are slightly different in conformation and performance, the latter tending to have more of the beef qualities of its ancestors. American and Canadian sires are now being used extensively in Europe to improve the original strains.

A Brown Swiss Cattle Breeders' Association was formed in New Zealand in 1973 with the objective of introducing the breed here. The first half-bred calves from imported semen were born in 1975. Since then pure-bred bulls have been imported from England, and semen of European and American origin has been used.

APPEARANCE

A strongly built, deep-bodied animal with a well-placed udder and very good conformation in legs and feet. Colours range from light brown through to grey, with darker shades in those colours also occurring.

Left: Brown Swiss bull.

BREED FEATURES

Quiet, docile temperament. Long productive life of up to 15 years, with high milk yield and protein and fat production. Hardy, and tolerant of a wide range of climatic conditions and altitudes. A dual-purpose breed suitable for beef or dairy production, and for cross-breeding. It has been used extensively to upgrade other breeds.

Breeding

Slightly later maturing than many other breeds. Good fertility and easy calving. Calf birth weight averages 43 kg. Good mothering. Calf weight gain is quite rapid at about 1 kg a day.

Milk production

Brown Swiss cows are slower than many other breeds to achieve their peak lactation, and may produce unspectacular results in their first year. Thereafter, however, their performance is very good and peak is reached at or after the sixth lactation, followed by a long productive life. Many cows will produce over 5000 litres of milk and 200 kg of milkfat in a lactation, with a high protein content that makes the milk ideal for cheese-making. Average milkfat content exceeds 4 per cent, with many cows exceeding 4.5 per cent.

Growth and carcass

Good live weight gain with large carcass at maturity.

Height and weight

Mature bull:	Average height at withers 145 cm	Average weight 1000 kg
Mature cow:	Average height at withers 134 cm	Average weight 720 kg

Right: Brown Swiss cow and calf.

Charolais   (B)

Developed in the Charolles district in central France, the ancestors of today's Charolais were used for draught and meat. Improvement through selective breeding began early in the 19th century, and the French Charolais herd book was established in 1864. In France the pure-bred Charolais population is currently nearing 2 million, and the breed has developed a worldwide reputation as a fast-growing and profitable producer of quality lean beef.

The breed entered North America in the 1930s, and about 1.5 million pure-bred animals are now registered there, making them one of the most popular beef breeds. The Charolais was introduced into Britain in 1962, and used primarily to improve the cross-bred beef calves produced from the dairy herd. However, some of the imported bulls produced difficult calvings and as a result many dairy producers lost confidence in the breed. During the last 30 years selection for easy-calving bloodlines has restored that confidence and it is currently the most popular beef breed, and for artificial insemination the second most popular beef sire.

British semen was imported into New Zealand in 1965. The New Zealand Charolais Cattle Society was formed in 1968, and during the next 13 years 61 bulls and 302 females were imported from the United Kingdom and Ireland, to form the basis of the Full French Charolais. In 1973 12 females were imported from France; progeny from these dams are kept on a separate register. Pure-bred (minimum 31/32) cattle have been produced through grading up, using a French Charolais sire over a base Angus, Friesian or Hereford cow. These 'New Zealand' Charolais are better suited to local production, having been selected for a smoother muscle pattern to assist ease of calving when mated to the smaller-framed dam breeds used in New Zealand.

APPEARANCE

Two distinct types of Charolais may be seen in herds, the American and the European. The former is longer in the leg and taller than the smaller and chunkier, more heavily muscled European animals. The British type lies somewhere between the two. The Charolais is large, long-bodied and heavily muscled. Coat colour is normally white or very light straw; some dark straw and red animals are throwbacks from grading-up and not generally recognised by national breed societies. Most animals are naturally horned, but polled animals have been developed and are becoming more common.

BREED FEATURES

A vigorous, hardy and thrifty breed with a high degree of feed efficiency and impressive average daily weight gain. Pink eye and eye cancer are practically non-existent.

Breeding

Some North American strains mature earlier than the European strains. Gestation of about 285 days is longer than that of many cattle breeds. Calf birth weight is about 41 kg, but there are wide variations between strains. Cows have very good mothering and milking ability, which is reflected in the fast growth rate of their calves. Charolais bulls are often used as a terminal sire for cross-breeding, and because the Charolais colour is dominant their progeny from Aberdeen-Angus cows are a distinctive white or silver colour while those from Herefords are known as 'khaki baldies'.

Growth and carcass

Weight gain of about 45 kg per month for the first 12 months is the standard, with many bulls maintaining this growth to 18 months or more. Heavy muscling in the loin and round produces a high percentage of prime cuts with a minimum of subcutaneous fat. Good carcass yield and lean, lightly marbled meat.

Height and weight

Mature bull: Average height at withers 148 cm   Average weight 1050 kg
Mature cow: Average height at withers 137 cm   Average weight 850 kg
North American strains are taller.

Below: Charolais calf, bull and cow.

Chianina .... Ⓑ

Originating in the Chiana Valley in Tuscany, in west-central Italy, the breed is now found in various parts of Italy. Like many Italian breeds it has a blood factor in common with the zebu (*Bos t. indicus*), and its ancestors were the large, white shorthorned cattle brought to Italy by the Etruscans more than 2000 years ago and used by the Romans for sacrificial purposes. The breed has four distinct types, the largest (making the Chianina the largest cattle breed in the world) being the Val di Chiana, which is found on the plains and hillsides of Arezzo and Siena. An official breeding programme was commenced in 1932 under the sponsorship of the Italian Government. The Chianina was still used for draught purposes in Italy until comparatively recently, but selection is now for beef. It is used throughout the world in cross-breeding programmes to help in the improvement of local cattle, and its heat-tolerance makes it particularly useful for areas with hot climates.

The Chianina arrived in New Zealand in the early to mid-1970s, but remains a minority breed. It has been cross-bred with the Aberdeen-Angus to produce the Chiangus.

APPEARANCE

A very large breed. Head comparatively long. Horns short, black on young animals, turning to a yellowish colour with black tips by 2 years of age. Very high shoulders with slight hump, long wide loins and sloping rump. Well-developed dewlap. Skin dark-pigmented with white hair, coat short and smooth. Tail switch, eyelids and hooves are black. At birth calves are tan-coloured, changing to white at 3–4 months of age.

Left: Chianina cow.

BREED FEATURES

Docile, powerful and possessing great stamina. The pigmented skin makes the Chianina comparatively heat-resistant.

Breeding

Despite its size there are few calving problems if bulls are carefully selected.

Growth and carcass

Carcass weight and quality is very good with a high proportion of lean meat and prime cuts.

Height and weight

Mature bull:
Average height at withers 172 cm. Individuals can reach 180 cm or more
Average weight 1250 kg. A world record set by 'Donetto' in 1955 was 1740 kg
Mature cow: Average height at withers 160 cm   Average weight 850 kg

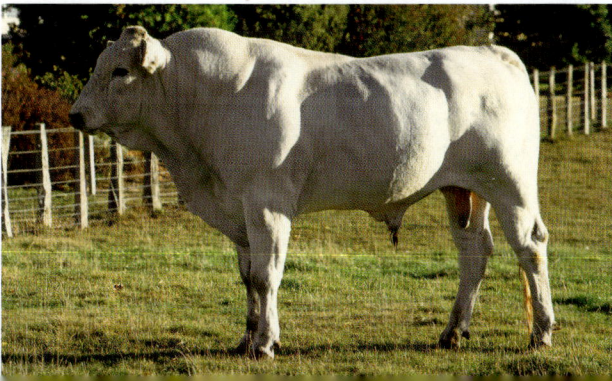

Above: Chianina calf.

Left: Chianina bull.

Devon or Red Devon .... Ⓑ

One of the oldest breeds of British cattle, the Devon's origins are unknown although it could have links with the Neolithic Longifrons (*Bos longifrons*). There are marked similarities with the French Salers, and during the 19th century Devon cattle were used to improve the Salers breed. The red coat and inherited heat tolerance suggest possible links with the red cattle of India or the North African red cattle which may have been brought over by the Phoenician traders to exchange for Cornish tin. We also know that the Romans used heavy red draught oxen for road building. Whatever its true origin, the breed was gradually standardised in the five counties of south-western England, particularly the Exmoor region of north Devon, and has been recognised as a distinctive breed for more than 300 years. The earliest written records indicate that the Devon was sent to Kerry in 1580.

Like most breeds it was originally used for draught as well as meat and milk, but it is now used for beef production. The first volume of Davy's Devon Herd Book was published in 1851, and the Devon Cattle Breeders' Society was formed in 1884.

In the mid-19th century in Britain the breed was second only to the Shorthorn in popularity. Later competition from Angus and Hereford cattle resulted in a decline in numbers, and this was compounded by the invasion of Continental breeds from the late 1960s. Increasing interest in cross-breeding programmes during the last two decades, problems with some of the introduced Continental breeds, and better promotion of the Devon have resulted in renewed interest in its grazing ability, low concentrate requirement, easy care and meat quality. Salers and Limousin bloodlines have been used in some breed improvement programmes.

The breed was first imported into New Zealand in 1838, being used mainly for draught and forming the basis of the bullock teams that were used in the early timber industry.

BREED FEATURES

A very docile, compact and muscular beef breed. Heat tolerant, the red coat colour aiding protection against the sun and the red eye pigmentation helping to prevent eye cancer. Hardy and robust, a very good forager, and adaptable to a wide range of soils, climates and environments.

Breeding

Early maturity and good fertility, with an 85–90 per cent calving rate under natural conditions. Easy calving due to small head of calf; good mothering. Calf birth weight averages 36 kg. Devon bulls are recommended as terminal sires for cross-breeding with dairy cows to produce high-quality dairy-beef calves.

Growth and carcass

Cows produce good milk yields for their suckling calves, whose good growth rate continues after weaning. Very good feed conversion with an ability to fatten on natural grass and good carcass yield. The hindquarters weigh heavier than the forequarters. Meat red and lean with good marbling, flavour and tenderness.

Height and weight

Mature bull:
Height not normally measured, but rump height of two outstanding bulls averaged 148 cm
Average weight 1050 kg
Mature cow: Average weight 700 kg

APPEARANCE

A medium-large strong-boned breed with a broad forehead and a deep, long and well-balanced body. Horns are placed at right-angles to the head and on most animals curve downward. Abundant coat of deep rich red mossy hair. Originally naturally horned, but a polled variety now exists which accounts for about 25 per cent of registrations.

Facing page: Devon cows and calves.
Right: Devon bull.

Dexter ⸺⸺⸺⸺⸺⸺⸺⸺⸺⸺⸺⸺⸺⸺ (D)(B)

The Dexter is a short-legged version of the Kerry, developed from ancient Celtic cattle that lived in Ireland in Neolithic times. The first report of a 'miniature' cattle breed was in 1776, although it was not until the 19th century that it was named the Dexter after the Mr Dexter who developed it. It is assumed that the Dexter arose either as a mutation from the Kerry or by later breeding with it. The breed became popular in Ireland during the late 19th century and a Dexter herd book was commenced there in 1887, but by the early 1900s so many Dexters had been sold to England that the breed was almost non-existent in its native country and the Irish herd book was closed prior to World War I.

In England the number of Dexter cattle gradually dwindled to the extent that when the Rare Breeds Survival Trust was established in 1973 (see page 41) the breed was classified as Rare and placed on the Trust's Priority list. Numbers increased again and the Dexter was upgraded to the status of a Minority breed. The breed is now undergoing transition to a mainstream cattle breed.

Dexter cattle have been exported to several countries, including South Africa, where it is called the Dexter-Kerry and has its own herd book.

Dexters were first imported into New Zealand in 1904, but the breed did not survive for long. Dexter semen was imported in 1979, and a New Zealand Dexter Breeders' Association formed in 1993. This was later named the Dexter Cattle Society New Zealand Incorporated. As a result of breeding-

Left: Dexter bull.
Right: Dexter cows.

BREED FEATURES

A very distinctive breed whose small size appeals to many breeders. Easily handled.

Breeding

There are two types, long-legged and short-legged; both of which are registered in New Zealand. In the UK some short-legged animals carried the chondrodysplasia gene, which occasionally resulted in the premature death of the calf. This problem now appears to have been solved.

Milk production

Dexter cows average about 2650 litres of good-quality milk with high milkfat and protein levels.

Growth and carcass

As beef animals, Dexter steers can finish off grass at 18–24 months of age at a live weight of up to 350 kg.

Height and weight

Mature bull:	Average height at withers 107 cm	Average weight 450 kg
Mature cow:	Average height at withers 100 cm	Average weight 325 kg

up programmes there are now more than 3000 Dexters of various grades of purity, but very few pure-bred animals.

APPEARANCE

The smallest of the modern cattle breeds. Coat colour black, red or dun. The bull's coat becomes curly over the forequarters in winter. Naturally horned.

Galloway ⓑ

It is generally agreed that the Scottish beef breeds owe their origins to cattle spreading from northern Europe around 2000 BC. The Galloway's ancestors were indigenous small-horned black cattle from which Scottish farmers started to select polled individuals, and this particular breed was developed in the Galloway region of south-west Scotland. Originally there were a variety of colours, but the first herd book established in 1877 only permitted the registration of black animals. Later that restriction appears to have been partly lifted, and prior to 1911 the dun-coloured animals that spontaneously occurred in black herds were apparently registered as Galloways. This assumption arises because at the breed association's AGM in 1921 it was agreed that the produce of Dun Galloway cattle entered in the herd book prior to 1911 be eligible for registration. The first animals to be entered with the (DUN) suffix after their pedigrees appeared in the 1923 herd book. Another colour variation, black with a white band around the body, is registered as a separate breed (see Belted Galloway, page 84).

In Scotland the Galloway was bred relatively low in weight to reduce damage to soil and pasture, but in North America larger-framed animals were developed and these bloodlines are now used where heavier animals are required. It is said that in America it was a Galloway bull that was used to introduce the polled factor into Hereford cattle.

Galloways were first imported into New Zealand in 1947 by farmers who were impressed by the breed's ability to produce high-quality beef from rough moorland grazing, and a breed society was formed in 1948. Competition from

Left: Dun Galloway cow and calf.

Right: Silver Galloway calf.

Far right: Galloway bull.

BREED FEATURES

Hardy and adaptable, a very good forager, tolerant of cold harsh climatic conditions and does not require housing in winter. Ideal for high-country stations, but also useful in less rigorous climates. Good conversion of rough feed into weight gain. Females have very strong mothering instincts and are very protective of new-born calves, which has led to a reputation, probably undeserved, for being temperamental.

Breeding

Easy calving. Calf birth weight about 36 kg. Cows have very good mothering and good milking ability, are long-lived and regular breeders, and will pass on these traits in cross-breeding programmes.

Growth and carcass

The thick double-hair coat results in less subcutaneous fat and a leaner carcass. The meat is well marbled.

Height and weight

Mature bull: Average height at withers 135 cm   Average weight 800 kg
Mature cow: Average height at withers 120 cm   Average weight 500 kg
North American bloodlines are taller and heavier.

European beef breeds during the last couple of decades has resulted in less interest being shown in the breed.

APPEARANCE

A medium-sized beef breed with a deeply rounded body and deep, well-sprung ribs. A unique shaggy double coat comprising an outer layer of soft wavy hair and a mossy undercoat. Coat colour black or dun. Red is also accepted in some countries, as is silver (white) with black or dun points. Naturally polled.

Gelbvieh ⓑ

Developed from a number of strains of local 'Red-Yellow Franconian' cattle in the Northern Bavarian region of Germany in the 19th century, the name literally translates as 'yellow cattle'. Originally used for meat and milk production, and as a draught animal, it is now generally regarded as a beef breed. A key feature has been careful selection for maternal and structural soundness as well as productivity.

The breed first arrived in New Zealand in 1976. The Gelbvieh Cattle Breeders' Society of New Zealand was formed in 1990. There are presently 35 members and approximately 1700 registered cattle in the breeding nucleus. Live cattle have been displayed at the New Zealand National Agricultural

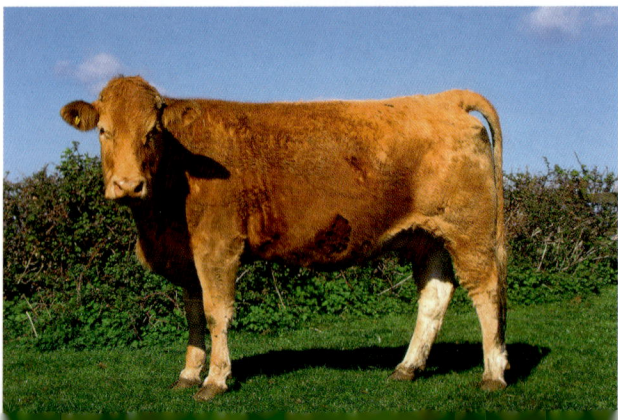

Above: Gelbvieh bull.

Left: Gelbvieh cow.

BREED FEATURES

Very docile temperament and good maternal traits make for easy management. Good conformation of udder is ideal for suckling calves. Fast growing, and reaches puberty earlier than most other beef breeds. Adapts well to extensive range of conditions and climates ranging from cold to subtropical. Recommended for cross-breeding with other beef breeds to improve performance of commercial breeding females, especially the first (F1) generation. Pigmented skin helps protect against eye cancer caused by ultraviolet light, an advantage when cross-breeding with white-faced breeds. In New Zealand the predominant market is cross-breeding over British beef breeds, but selected males can also be used as terminal sires for dairy breeds. Bulls are easily managed.

Breeding

High fertility. Heifer and bull calves should be separated by 6–8 weeks old to avoid juvenile pregnancies, as heifer calves are capable of conceiving at 3½ months. Pure-bred females calve easily, but care needs to be taken in the choice of sire when cross-breeding with maiden heifers of smaller breeds. Good mothering. Calf birth weight averages 39 kg.

Milk production

Careful selection has produced a sound udder with well-placed teats. Good milk production, with average milkfat level of 4 per cent, and protein level of 3.45 per cent.

Growth and carcass

Above average pre-weaning and very good post-weaning growth. Very good feed conversion on good pasture or grain, and better than average on poor pastures. Meat lean while retaining some marbling.

Height and weight

Mature bull:	Average height 148 cm	Average weight 1000 kg (up to 1300 kg)
Mature cow:	Average height 135 cm	Average weight 725 kg (up to 900 kg)

Fieldays, near Hamilton, every year since 1992. Many of the cattle displayed have been unled commercial examples. This has helped demonstrate their excellent temperament, and increased the breed's profile.

APPEARANCE

A medium-large, long-bodied, muscular breed. Coat colour reddish gold to russet coloured, hair straight and of medium length. Black-coated strains also exist within the breed, and are increasingly popular in the US market for cross-breeding with black Angus.

Guernsey .... Ⓓ

The breed originated on the island of Guernsey in the Channel Islands, and is thought to have been developed as a result of the combination of two ancient French breeds. Around AD 960 a group of monks from St Michel du Vale in Brittany founded a colony on the island. To provide the milk, butter and cheese that were an important part of their diet they brought with them cattle which may have been the now rare breed called the Froment du Leon. About a hundred years later another group of monks arrived from Cherbourg, bringing with them the Normandy breed called the Isigny, and from the resulting cross was developed the Guernsey. By the early 18th century the breed was already becoming well known for its high-quality milk, and landowners close to Britain's south-coast ports had begun to import Guernsey cattle. These were initially used to establish dairy herds for the 'home farm', but later larger commercial herds were established. Between 1814 and 1842 a series of laws were passed by the Royal Court of Guernsey that restricted or prohibited importation of cattle to the island in order to protect the purity of the breed. Breed improvement and standardisation continued for the next 50 or so years, with much of the best breeding stock being sent to the United States, where an American Guernsey Cattle Club was finally established in 1877. The first Guernsey Island herd book was published in 1878, and the English Guernsey Cattle Society was formed in 1884.

A population of about 2500 animals is still maintained on the island of Guernsey, but like many dairy breeds the Guernsey has declined in popularity because of competition from the Holstein-Friesian and a demand for milk with a lower milkfat content. In recent years semen from the United States has been used to improve the British stock, resulting in larger cows with less flesh cover. In Britain a number of herds still produce milk for the manufacture of ice cream, and in the south of England Guernsey herds are currently supplying a niche market for a range of high-quality dairy products.

BREED FEATURES

Very docile temperament and easily managed.

Breeding
Easy calving, calf birth weight about 36 kg. Very good longevity.

Milk production
Cows routinely produce high-quality milk over five lactations or more. The milk is noted for its high milkfat content (averaging 4.7 per cent), and its rich golden-yellow colour.

Height and weight

Mature bull:	Average height at withers 138 cm	Average weight 800 kg
Mature cow:	Average height at withers 133 cm	Average weight 550 kg

Despite some early imports, at the end of 2007 there were only 30–40 purebred animals in New Zealand. Some grading-up programmes are being undertaken using semen from American bulls.

APPEARANCE

A small, fine-boned dairy breed, of larger build than the Jersey. Head long and narrow with a cream-coloured muzzle. Hindquarters comparatively narrow. The skin has a yellowish tinge, and the coat colour may vary from a yellow-fawn to brown or red, with or without white markings on the underline and legs. The hooves are amber, the tail switch white. Naturally horned.

Facing page: Guernsey cow. *Above*: (*left*) Young Guernsey bull. (*right*) Guernsey bull.

Hereford .... Ⓑ

The exact origin of the Hereford is uncertain, but it is generally agreed that it was founded on the draught ox descended from the small red cattle of Roman Britain, and from a large white Welsh breed that was once very common along the border of England and Wales. The now extinct Glamorgan breed may also have been involved in the Hereford's evolution. Whatever its origin, the Hereford was certainly one of the first English cattle breeds to be developed, and as early as 1627 the cattle from that English county were in demand in all parts of the country as yoke oxen, and had an excellent reputation for their milk and beef production. Descriptions in literature of the day prove that the classic Hereford type, with its red and white markings and white face, had already been fixed by the year 1788. The white face is possibly the result of cross-breeding with Flanders cattle introduced into England during the second half of the 17th century. Benjamin Tomkins 'the Elder' (1714–1789) and his son Benjamin Tomkins 'the Younger' are credited with establishing the foundation for the present line during the second half of the 18th century, and in 1775 Herefords were introduced into County Westmeath, Ireland. The first recorded sale of the breed was in 1795. The first volume of the English Hereford herd book was published in 1846, and the Hereford Herd Book Society was founded in 1878.

The first Hereford imports into New Zealand were a bull and a cow, which arrived around 1868. In 1875 a bull and three heifers were imported, and many of the registered cows in the country today are descendants of one of those heifers, called Amethyst. In 1886 the first New Zealand herd book for cattle other than Shorthorns was published; it listed 58 bulls and 317 cows of the Hereford breed. Formed in 1896, the New Zealand Hereford Cattle Breeders' Association published its first herd book three years later.

Left: Hereford bull.

BREED FEATURES

Docile. Hardy and adaptable to a wide range of environmental and climatic conditions. A very good forager on pasture, but also suited to intensive conditions. A well-proven and very popular beef breed.

Breeding

High fertility. Gestation comparatively short at approximately 280 days. Easy calving. Calf birth weight averages about 38 kg. High percentages of live births and calves weaned. Bulls and cows both have very good longevity, and are commonly used in cross-breeding programmes.

Growth and carcass

Very good feed conversion. Medium maturity. High-quality carcass.

Height and weight

'Modern' strains may be significantly taller than 'traditional' strains. The measurements below are from data on show cattle (Royal Show 1993) and likely to be higher than the general average.

Mature bull:	Average height at rump 153 cm	Average weight 1147 kg
Mature cow:	Average height at rump 141 cm	Average weight 700 kg

More than 5 million pedigree Herefords (including the polled variety) now exist in over 50 countries, and Hereford bulls are widely used as crossing sires on commercial cattle and indigenous breeds. Because of the way the breed has been developed in various countries, there are quite significant differences between the 'modern' and 'traditional' Hereford strains, the latter being smaller and more compact. The Hereford Herd Book Society now keeps a separate register for animals of pure, traditional English descent.

APPEARANCE

A medium-sized, stocky beef breed. Colour deep red with a creamy-white face, chest, flanks and leg points. Originally naturally horned, but the Poll Hereford (see page 110) is now becoming the most common variety.

Right: Hereford cow and calf.

Poll Hereford   (B)

The Poll Hereford originated in the United States. In 1894 on the Iowa property of C.T. Mercer the chance mating of a Red Poll bull and a Hereford cow produced a polled calf. This stimulated interest in a polled variety and around 1899 five mid-western breeders decided to select for a polled strain. It is reported that a Galloway bull was selected to introduce the original polled factor. In 1901 a breeding group of polled animals was set up, and an American Polled Hereford Cattle Club, the forerunner of today's American Polled Hereford Association, was formed.

Attempts to develop a similar programme in Britain using British stock were not very successful, and it was only after American bloodlines were introduced that it finally got under way. The polled variety is now becoming quite common.

The first Poll Herefords in New Zealand were derived from American bloodlines, and were first registered in 1928. Both polled and horned varieties are registered by the New Zealand Hereford Association, and polled animals now account for around 75 per cent of all cattle in the herd book.

Right: Poll Hereford cow and calf.

MINIATURE HEREFORD

Height and weight

Mature bull:	Average height at hips about 110–125 cm	Average weight 650 kg
Mature cow:	Average height at hips about 100–120 cm	Average weight 420 kg

Miniature Hereford   (B)

Miniature Herefords were first introduced into New Zealand in 1997. Embryos were imported from Canada and live animals from Australia to establish the first breeding stock in New Zealand. Miniature Herefords originated in Texas and were bred by the Largent family, who have been breeding top show Hereford cattle since the early 1930s.

Since then the interest in these smaller chunky Herefords has grown, and the appeal of easy handling and less pasture damage has given smaller acreage farmers who love Herefords the chance to have the breed they prefer to suit the size of their farm. With 135,000 lifestyle farms in New Zealand, and the number growing fast, the Hereford breed can now fit all size farms and breeders.

Miniature Herefords are pure Hereford and have to pass the strict criteria of the New Zealand Hereford Association. Until 2007 they were registered with the Australia and New Zealand Miniature Hereford Cattle Association, but that year a group of Miniature Hereford breeders formed the New Zealand Miniature Hereford Breeders' Group. This runs in conjunction with the New Zealand Hereford Association and has rules and regulations that maintain strict criteria on eligibility to register an animal as a Miniature Hereford. The New Zealand Hereford Association and the New Zealand Miniature Hereford Breeders' Group will work together to maintain the quality and purity of the breed. At the end of 2007 the Miniature Hereford Breeders' Group had 21 breeders with around 150 animals.

Left: Miniature Hereford.

Highland .... (B)

Supposedly descended from ancient Celtic longhorned cattle, and already known in the 12th century, the breed was developed on the islands and mainland of western Scotland. It gained prominence in the 19th century with the development of a trade in beef animals from Scotland to English markets. The Highland Cattle Society was formed in 1884 and the first herd book (bulls) was published in 1885. Most of the early animals registered were black. Breed numbers declined after World War II, but since the late 1970s numbers have markedly increased and Highland Cattle Societies have been formed in many countries.

Three cows and a bull were imported into New Zealand from Scotland in 1973, and one cow from Canada in 1979. Most existing Highland cattle originated from cross-breeding programmes, using semen imported mostly from Scotland and Australia. The New Zealand Highland Cattle Society was registered in October 1993. In May 2006 there were 628 full-bloods, 1223 pure-breds and 2254 graded animals. At the end of 2007 there were 425 members and 529 registered breeders or folds. Some members are now importing embryos to bring new bloodlines into the country and there is a good selection of semen available from the United States, the United Kingdom, Australia and Sweden.

APPEARANCE
A comparatively small, long-bodied breed. Coat long and shaggy, with coarse hair hanging down over the eyes. The coat grows extremely thick in winter. Colour may range from cream or yellow-red through to brown or black. Horns long, spreading horizontally and usually rising at the tips.

Left: Highland cow.

Right: Highland calf.

Far right: Highland bull.

BREED FEATURES

An extremely hardy and versatile breed. Its prime feature is its ability to thrive on poor mountain land with a high annual rainfall and a cold wet environment, but it is also playing a role on better pastures in more equable climates. Highland cattle are popular for lifestyle blocks as they are easy-care, good foragers and good mothers. They handle extreme weather conditions and are an excellent-quality beef.

Breeding

Calves are born extremely hardy, weighing an average of less than 40 kg. Cows have very good longevity, a significant number still calving at 18 years of age. As well as being used to produce pure-bred calves, Highland cows have also been used for cross-breeding with Shorthorn-type cattle to produce offspring with good hybrid vigour. More recently very good results have been obtained in cross-breeding programmes with some of the Continental beef breeds.

Growth and carcass

Although slow-maturing the Highland is able to convert poor forage into a very marketable carcass averaging 52 per cent yield. Meat is lean, well marbled and of good flavour. Some New Zealand dairy farmers are choosing to use Highland bulls over their dairy heifers as they breed a smaller calf but with a good beef yield.

Height and weight

2-year-old bull:	Average height across the hips 119 cm	Average weight 675 kg
Mature cow:	Average height across the hips 109 cm	Average weight 450 kg

Holstein-Friesian (Dutch Friesian; Holstein) .... Ⓓ

The Dutch Friesian and American Holstein form the basis of the Holstein-Friesian, and the three breeds together make up a worldwide population of dairy cattle which are often called by the simple name of Friesian.

The exact origin of the Dutch Friesian or Dutch Black Pied is obscure. During the 18th century small black and white cattle were brought from Jutland into northern Holland and Friesland to replace animals lost through flooding and disease. Crossed with the original Dutch cattle they became the basis of the Dutch Friesian breed. At that time cattle were both black-pied and red-pied, and maintained separately, but with the establishment of a Netherlands herd book in 1873 and the Friesland herd book in 1879 a preference for the black and white cattle, particularly in America, led to their becoming dominant.

The Holstein breed originated in America following the importation of Dutch Friesian cattle during the mid-19th century. The breeder, W.W. Chenery, used the word 'Holstein' (possibly as a corruption of Holland) and following further importations a Holstein-Friesian herd book was published in 1885.

In Holland and America breeders then followed different paths, the Americans developing the original large Dutch Friesian into an even larger type, whose name was officially shortened to Holstein in 1978. The Dutch selected for a smaller, more compact dairy-beef animal which finally fell from favour with changing world markets during the 1970s, and was then improved by the addition of Holstein blood.

During the last few decades most breeders throughout the world have used Holstein blood to improve the size and performance of their original Dutch Friesian stock, so that today's 'Friesians' are almost always a product of the two bloodlines; hence the name Holstein-Friesian.

The Dutch Friesian was first imported into Britain from Holland in 1914, and following a period in quarantine 39 bulls and 20 heifers were put up for sale. The Holstein first arrived in Britain in 1946, when 220 animals were brought in from Canada. From 1956 Livestock Quality Controls stopped the importation of semen into Britain from the United States, greatly slowing down the 'Holsteinisation' of the breed. In the mid-1980s US bloodlines once again became available, and since then Holstein blood has been progressively introduced into the UK Friesian population. There are separate breed societies for the Holstein and the Holstein-Friesian in the United Kingdom.

BREED FEATURES

Adaptable to a wide range of environmental conditions. Exceptional ability to convert grass and roughage into milk and protein. High milk protein–fat ratio. Renowned for its strength of constitution and longevity.

Breeding
High fertility. Easy calving, calf birth weight about 41 kg. Good mothering, excellent milk production. Very good longevity.

Milk production
Compared with other dairy breeds the Holstein-Friesian has the highest average annual milk production, and although its milkfat and protein percentages are usually lower, the total weights of protein and milkfat produced during a lactation usually exceed those of the other breeds. The high ratio of milk protein to milkfat averages 0.82:1, and makes the milk ideal for cheese production.

Growth and carcass
Friesian cows not required as breeders for herd replacement are usually crossed with a terminal sire from a beef breed, their large frame helping to produce very good dairy-beef calves. In addition to being reared for beef production, Hereford × Friesian females are commonly used as recipient cows for embryo transplants (see page 38).

Height and weight
Figures for height and weight vary according to bloodlines, animals with a high proportion of Holstein blood being significantly taller and heavier.
Mature bull: Average height at withers 150 cm   Average weight 1100 kg
Mature cow: Average height at withers 135 cm   Average weight 700 kg

Above: Holstein-Friesian bull.

The first importation of Dutch Friesians into New Zealand was from Holland in 1884, but the spread of the breed was slow until the early part of the 20th century when large numbers of cattle were imported, particularly from Australia, Canada, Britain and the United States. A New Zealand Friesian Association was formed in 1910, and the breed became relatively popular. After World War II, however, there was increasing emphasis on milkfat production for the manufacture of dairy products and for many years the Jersey breed dominated the dairy industry. In recent years the Holstein-Friesian has become dominant, and approximately 55 per cent of inseminations are with semen from Holstein-Friesian bulls. These are mainly New Zealand bred, but bloodlines are also derived from Britain, Canada and the United States. The introduction of cross-bred sires (predominantly Holstein Friesian × Jersey) has reduced the number of Holstein-Friesian inseminations.

APPEARANCE

A large dairy breed with a deep, long body. Udder of moderate length, width and depth, slightly quartered, soft and pliable, and well veined. Teats of medium size, cylindrical, and placed squarely in the centre of each quarter. Coat colour black and white in various proportions.

Above: Holstein-Friesian cow.   *Right*: Holstein-Friesian calf.

Jersey .... Ⓓ

Although it is named after the Channel Island on which it was developed, the origin of the Jersey breed is unknown. There is evidence that cattle were originally brought to the island by Neolithic settlers, and a skull of the Neolithic Longifrons (*Bos longifrons*) was unearthed there, but the breed probably evolved from French cattle taken to Jersey from Normandy and Brittany. On the island the breed has been kept pure through a series of import regulations, the first of which was passed in 1763, followed by a more stringent one in 1789. Later amendments allowed importation for slaughter or re-export, but strict precautions were taken to keep the breed pure and disease-free. The Jersey Agricultural and Horticultural Society was established in 1833 with the aim of improving cattle breeding and general agriculture, and a year later received the patronage of King William IV. It has had royal patronage ever since. The Jersey herd book was founded in 1866.

In New Zealand the Jersey attained its strongest position, being ideally suited to the climate and the accent on grassland farming. The first Jerseys in New Zealand were a bull and two cows imported from the island of Jersey in 1862. Other importations followed, many of the females being cows brought on sailing ships to provide fresh milk during the voyage. The New Zealand Jersey Cattle Breeders' Association was formed in 1902, but the breed made comparatively slow progress and dual-purpose breeds predominated until the introduction of refrigerated shipping made possible the export of dairy products to Britain and Europe. After that the Jersey rapidly rose to number one dairy breed. However, in recent years breed numbers have declined and the prime position has been taken over by the Holstein-Friesian.

APPEARANCE

A small, fine-boned dairy breed. Characteristic short head and dished face. Skin fine and covered with soft fine hair. Outside its native island most breeders select for a fawn coloration, but in Jersey itself a variety of colours are found. Muzzle dark, surrounded by a ring of light hair. Naturally horned, horns inward curving.

BREED FEATURES

Docile temperament and easily handled. Light weight reduces damage to grassland. Efficient feed conversion. Very good milk production and superior fat and protein levels, especially when assessed per kilogram of body weight and per hectare of pasture.

Breeding

Early maturing. Very good fertility. Easy calving, calf birth weight about 34 kg. A worldwide Jersey Genetic Recovery and Classification programme is in place to ensure the maximum retention and use of desirable genes.

Milk production

Individual cows can produce more than 7500 litres of milk, 450 kg of milkfat and 300 kg of protein per lactation. In New Zealand many farmers target production from a given area of land rather than per cow, and Jersey farms can average 700 kg of milkfat and 450 kg of protein per hectare.

Height and weight

Mature bull:	Average height at withers 124 cm	Average weight 690 kg
Mature cow:	Average height at withers 120 cm	Average weight 390 kg

Facing page: Jersey cow. *Above*: Jersey bull.

Limousin .... (B)

Located in the harsh environment of the Massif Central, this very old French breed was originally selected during the 17th century for use as a draught animal. It was also required to exist on sparse vegetation and forage widely to survive, and this combination of strength and ability to grow on poor diets has been passed on through the generations. Breeding improvements began around 1840, and in 1886 a Limousin herd book was started. The Limousin gradually became recognised for its quality meat production, and in 1935 an elite herd book was started, the first for any beef breed. During the last 30 years in particular, pedigree Limousins have flourished and have been exported to about 60 countries, thriving in conditions as dissimilar as the cold of northern Canada and the tropical heat of northern Queensland. Limousin beef is marketed under the international brand name 'Limousine Gourmet'.

The breed entered Britain with the arrival of 178 animals in February 1971, and was initially used mainly for cross-breeding with dairy cows for the production of dairy beef. Since then the Limousin has become well established, and in 1995 approximately 11,000 pedigree animals were registered, with about 300,000 inseminations being carried out annually, mainly in dairy herds. The most striking change in recent years has been the increasing use of the Limousin for the insemination of Friesian cows to produce Limousin × Friesian offspring, the female calves being grown on for use as suckler cows in beef suckler herds. Approximately 150,000 such Limousin-bull females are available annually, and this particular breed cross has largely replaced the Hereford × Friesian in the suckler cow sector of the UK market. The Limousin × Friesian suckler cow is often crossed back to a Limousin bull. In Scotland the Limousin × Angus is performing well and can be bred sooner than traditional crosses.

The New Zealand Limousin Cattle Breeders' Society was formed in 1973. Ten heifers were imported from France the following year, and there has been increasing interest in the breed.

APPEARANCE

A medium-sized, solidly built and heavily muscled beef breed with a comparatively long body and fine bones. Colour normally a rich golden-apricot, although black Limousins are now being bred. Naturally horned, but a polled variety also exists.

BREED FEATURES

Hardy and adaptable to a wide variety of environmental conditions, and an excellent forager. Above average feed conversion rate.

Breeding

High fertility. Easy calving with 90–95 per cent live calves on the ground. Calf birth weight about 40 kg. Fast-growing. Recommended for cross-breeding to improve performance.

Growth and carcass

Reaches slaughter age earlier than most European breeds. Very good carcass yield with high meat to bone ratio, little fat and minimum waste. Meat finely textured, tender, and low in saturated fats and cholesterol.

Height and weight

Mature bull:	Average height at withers 140 cm	Average weight 975 kg
Mature cow:	Average height at withers 137 cm	Average weight 600 kg

Left: Limousin bulls.
Below left: Limousin cow.
Below: Limousin calf.

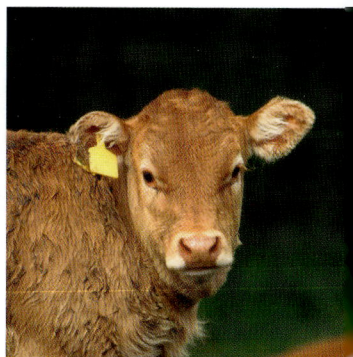

Longhorn .... (B)

The exact origin of the Longhorn is not known, but it is certainly an ancient breed. In the mid-18th century it became the first breed to be improved through the technique of inbreeding pioneered by Robert Bakewell of Leicestershire, England, who developed longhorned cattle native to central England into the Dishley or Leicestershire Longhorn. He selected for quick growth and heavy hindquarters, characteristics that soon saw it become the most popular cattle breed in Britain. Not only a beef producer, it was used extensively for draught and also for milk production. The long horns became a problem once farmers started to 'loose house' their beef cattle. Working oxen were replaced first by horses and later by tractors, and specialist dairy breeds took over the role of milk production. As a result the Longhorn declined until in 1965 there were only four major herds left. The breed was classified as Rare by the Rare Breeds Survival Trust when it formed in 1973, but since then interest in the breed has revived. A number of factors have helped the breed to recover. These include an active breed society promoting the breed and organising sales and shows; the creation of a semen bank; niche marketing of meat and hides; and the use of the breed in conservation grazing. The breed now has Minority status in the United Kingdom.

There are small numbers of Longhorn cattle in New Zealand.

APPEARANCE
A large breed. Colour red-brown or brindled, line-backed. Very long horns that tend to curve in unusual shapes, often curling inward very close to the face.

BREED FEATURES

Hardy, good maternal qualities, very suitable for cross-breeding.

Breeding
Easy calving. The calf is born relatively small, averaging about 31 kg, but thereafter has a comparatively rapid growth rate.

Height and weight

Mature bull:	Average height at withers 150 cm	Average weight 950 kg
Mature cow:	Average height at withers 135 cm	Average weight 650 kg

Above: Longhorn cow.
Far left: Longhorn ox.
Left: Longhorn calf.
Right: Longhorn cow with inward-curling horns.

Luing .... (B)

This hardy beef breed was established on the island of Luing off the west coast of Scotland by the Cadzow brothers, who commenced a breeding programme in 1947 by selecting their best first-cross Beef Shorthorn-Highland heifers and breeding them to a Beef Shorthorn bull. Through a combination of inbreeding and line breeding the breed finally became established and was officially recognised in 1965. The Luing Cattle Society was formed in 1966 and the herd book started.

In 1984 a special register was established for Simmental × Luing ('Sim-Luing') heifers from registered Simmental sires and pedigree and Appendix A Luing dams. The register is operated by the Luing Cattle Society in conjunction with the British Simmental Cattle Society, and all animals accepted into it have a special 'SIM/LNG' ear tag to identify them.

A New Zealand Luing Society was formed in 1975, but went into recess in 1993 through lack of membership. Remaining members maintain links with the UK Society.

APPEARANCE
A medium-sized, ruggedly built breed. Coat comparatively long and shaggy, colours variable but usually ranging from red-brown to light brown. Naturally horned, but some polled herds have been developed.

Left: Luing cow.
Right: Luing bull.

BREED FEATURES

Docile and hardy, and ideally suited to cold, wet, hill and upland environments. An excellent forager and roughage converter. Very long-lived.

Breeding

High fertility. Regular breeding. Easy calving, with calf birth weight of about 43 kg and excellent survival rate. Good mothering and milking qualities. It is much used for cross-breeding with other cattle breeds: for example, Simmental × Luing (Sim-Luing) females are highly suitable for suckler herds on lower ground, while Luing bulls crossed with Friesian cows produce much sought-after females which in turn are crossed mainly to Continental sires such as the Charolais. Another useful cross is the Luing × Salers.

Growth and carcass

Although cows are often small compared with other beef breeds, calf growth rate is fast, and it is claimed that the Luing has the highest recorded ratio of calf weaning weight to cow weight of any cattle breed. Carcass quality and yield are good, with little waste and a product suitable for today's markets.

Height and weight

Mature bull:	Average height at withers 140 cm	Average weight 1130 kg
Mature cow:	Average height at withers 132 cm	Average weight 700 kg

Maine-Anjou .... Ⓑ

During the early 19th century a breed called the Mancelle existed in the Haut-Anjou area of north-western France. Local breeders imported Shorthorned Durham cattle from Britain (see also Shorthorn, page 152) and the resulting cross formed the basis for the Maine-Anjou, which was developed as a specific breed between 1839 and 1860. There are now about 650,000 pure-bred animals in the area, and the breed is exciting interest outside its native country. Originally used for draught as well as meat and milk production, it later became a dual-purpose dairy and beef breed. Although cows are still good milk producers the breed is now used principally for meat production.

The breed first arrived in Britain from France in 1972, and in 1973 one of the cows produced a heifer calf, the first Maine-Anjou to be born in England.

The first Maine-Anjou semen was imported into Australia and New Zealand from Britain and Canada in 1973, and the first live Maine-Anjou cattle were imported into Australia from New Zealand in 1976.

Black Maine Ⓑ

Similar in most respects to the Maine-Anjou, the black-coated variety is now becoming very popular, especially in Canada where it is being selected for longer body and leg length. The black colour makes it ideal for crossing with other black breeds such as Aberdeen-Angus to produce all-black beef animals that are taller, longer and more docile, and can be marketed at any age from 15 months to 3 years.

APPEARANCE

A comparatively large and long-bodied breed with a wide forehead, deep wide chest and well-developed hindquarters. Coat colour dark red with small white patches, solid red around the eyes. Light-coloured muzzle. The Black Maine may be solid black or have varying amounts of white markings, particularly on the tail, lower legs or under the belly. The breed is naturally horned, the horns arching forward. A polled variety now exists and is gaining in popularity, particularly with Black Maine breeders.

Right: Maine-Anjou cow and calf.
Far right: Black Maine heifer.

BREED FEATURES

A docile temperament makes for easy handling. The traditional red colour results from a double recessive red gene, and this makes it compatible for cross-breeding with other beef breeds because the resulting cross takes on the coat colour of the other breed: e.g. a Maine-Anjou × Hereford calf has Hereford markings. A proven and hardy beef breed with good feed efficiency and growth rate.

Breeding

Comparatively early puberty and good fertility. More than 93 per cent of calvings are classified as easy. Calf birth weight averages 45 kg, but many breeders are now selecting for birth weights under 40 kg. Good mothering qualities. Bulls produce very good results when used as terminal sires in cross-breeding programmes with other beef breeds such as Charolais and Hereford.

Milk production

Although now principally a beef breed, overall lactation yields are good, averaging 2970 litres at 3.7 per cent milkfat with yields of up to 4500 litres at 4 per cent milkfat not unusual.

Growth and carcass

Cows have a good milk yield, so their calves have a high growth rate. This continues after weaning with excellent feed conversion and good carcass yield. Meat is of good quality, lean and tender, finely grained and uniformly marbled.

Height and weight

Mature bull:	Average height at withers 152 cm	Average weight 1250 kg
Mature cow:	Average height at withers 142 cm	Average weight 900 kg

Mandalong Special .... (B)

This Australian breed was developed over a period of more than 18 years by Rick Pisaturo, of the Mandalong Stud, St Marys, New South Wales, who wanted to find out if the best attributes of certain breeds could be combined. He selected Brahman for its hardiness and production of a small but vigorous calf, Shorthorn and Charolais to produce a well-muscled carcass with adequate fat cover, British White which as a dual-purpose breed would introduce milking ability plus beef attributes, and Chianina to maintain size once the initial breeding programme was complete and four generations of stability had been obtained. As a result the Mandalong Special is composed of 62.5 per cent Continental breeds, 18.75 per cent British breeds and 18.75 per cent Brahman.

The first importation into New Zealand was by embryo transfer in December 1990, with the first calves being born in September 1991.

APPEARANCE
A comparatively large, muscular breed. Colour usually various shades of gold, described as gold, light gold and dark gold. Slight *Bos t. indicus* characteristics such as a small hump in males and a fold of loose skin under the belly.

BREED FEATURES

A custom-built beef breed that displays easy calving, exceptionally rapid growth and a high-yielding carcass.

Breeding

Early maturity, good fertility, easy calving, good milk production. Calf birth weight approximately 36 kg. Calf vigorous and hardy.

Growth and carcass

Rapid growth rate with very good feed conversion and high carcass yield.

Height and weight

Mature bull:	Average height at withers 160 cm	Average weight 1200 kg
Mature cow:	Average height at withers 150 cm	Average weight 850 kg

Above: Mandalong Special bull.
Left: Mandalong Special cow.
Right: Mandalong Special calf.

Marchigiana ·········· Ⓑ

In the 5th century, after the fall of the Roman Empire, barbarians settled along the north-eastern Italian Adriatic coast in the region of Marche. With them they brought grey-white cattle, which are believed to be the ancestors of the Marchigiana. The breed was improved around the middle of the 19th century by crossing with Chianina and Romagnola cattle, plus other local stock. Not as tall as the Chianina, the Marchigiana has medium-sized bones and was formerly used for draught as well as beef production. A herd book was established by 1933 and the breed now accounts for almost 10 per cent of the Italian cattle population.

Marchigianas were first imported into Britain in 1974. The principal herd is located on the Penllyn Estate in South Wales. Marchigiana bulls are mainly used as terminal sires on British and other Continental breeds.

The breed first arrived in Australia and New Zealand in the early 1970s. Since then semen and embryos have been imported from Canada to ensure the highest quality in the bloodlines.

APPEARANCE
A large, long-bodied and muscular beef breed. Head short and wide. Horns of medium size, varying in colour. Skin black-pigmented. Adult colour light grey or white with black muzzle.

Above: Marchigiana bull bellowing. *Right*: Marchigiana cows and calves.

BREED FEATURES

Docile temperament. Can withstand extreme climatic changes, an excellent forager, and thrives on difficult hill country and poor-quality pasture.

Breeding
Easy calving. Calf birth weight averages 40 kg. Good longevity.

Growth and carcass
Cows have good milk production and young animals have a rapid growth rate, reaching up to 360 kg at 6 months of age.

Height and weight
Mature bull:	Average height at withers 160 cm	Average weight 1320 kg
Mature cow:	Average height at withers 145 cm	Average weight 720 kg

Meuse Rhine Issel .... (D)(B)

This red-and-white (red-pied) dual-purpose breed was developed in the south-east and east of the Netherlands, in the catchment areas of the Meuse, Rhine and Issel rivers. It is also referred to as the Meuse Rhine Yssel. The Dutch herd book was formed in 1874, and the Meuse Rhine Issel now constitutes about one-third of the national dairy herd in Holland. A similar breed was also developed in Germany, where it is known as the Rotbunt. German Breed Societies were formed in 1900. Between 1920 and 1950 Belgium, Luxembourg, France and Denmark founded their own red and white dual-purpose breeds, based on Dutch and German stock.

Meuse Rhine Issel cattle were first imported into the United Kingdom and Ireland in the early 1970s as part of a breeding programme to improve the Shorthorn. Some breeders found the Meuse Rhine Issel more suitable and phased out their Shorthorn stock. The breed is now well established.

In 1975 a number of full-blooded Meuse Rhine Issel heifers were imported into New Zealand, but it was not until 1990 that a Meuse Rhine Issel Cattle Society was formed. The Meuse Rhine Issel remains a minority breed.

APPEARANCE

A solidly built, medium-sized breed. Good length and depth to the body, with a slightly arched back. Good and evenly divided musculature, particularly in the upper and hind quarters. Colour red and white.

Left: Meuse Rhine Issel cow and calf.

Right: Meuse Rhine Issel bull.

BREED FEATURES

Docile temperament, sound leg and foot structure, hardy and adaptable, reliable breeding, good performance on high-roughage diets, very good milk production with high fat and protein percentages, very good bull beef quality. A good dual-purpose dairy breed.

Breeding

Good fertility, comparatively short gestation period, easy calving. Calf birth weight about 43 kg. Calves born active and lively. Cross-breeding with Holstein-Friesian produces a black and white calf.

Milk production

The breed is noted for its production of large quantities of milk with high milkfat and protein levels. European statistics show that average milk yields are about 6000 kg, with milkfat over 4.3 per cent and protein close to 3.5 per cent. The protein has a high kappa casein-B content, ideal for cheese-making. Individual yields of around 8000 kg at 4 per cent protein are not uncommon.

Growth and carcass

Growth rate and feed conversion compare favourably with commonly used beef breed–dairy breed crosses, as do carcass yield and meat quality.

Height and weight

Mature bull:	Average height at rump 143 cm	Average weight 1050 kg
Mature cow:	Average height at rump 132 cm	Average weight 675 kg

Murray Grey ⬚⬚⬚⬚⬚⬚⬚⬚⬚⬚⬚⬚⬚⬚⬚⬚⬚⬚⬚⬚⬚⬚⬚⬚⬚⬚⬚⬚⬚⬚⬚⬚⬚⬚ (B)

This Australian beef breed originated at Thologolong, a property on the Murray River in New South Wales. A Shorthorn cow purchased in the early 1900s by Mr Sutherland was mated 12 years in succession to Aberdeen-Angus bulls, and each year gave birth to a grey calf. Mrs Edna Sutherland insisted that these 'mulberry' calves be kept although to Mr Sutherland their colour was an embarrassment. During the late 1920s Helen Player, who married Mr Sutherland's son Keith, expressed a wish to own the cattle, which she called 'Goldens', and her opportunity came at a dispersal sale in 1932 when she and her husband bought eight animals and 1000 acres of land around the old Sutherland homestead. She set about selecting for the Goldens which, when crossed with Angus bulls, continued to throw grey or dark grey calves. In 1940 a neighbouring farmer, Cleaver Gadd, bought one of the grey-coloured bulls, and his son Mervyn established a breeding programme using it over Angus heifers; 60–70 per cent of the offspring were grey. At that stage he had no thought of creating a new breed, but liked the cattle because they grew quickly and had high-quality meat. Helen Sutherland also continued with her breeding, but in 1945 a severe drought left her with only eight animals. In 1952 bush fires destroyed all the grass at the Gadd property, but rather than send the grey cattle for slaughter Mervyn shipped them 650 km south on agistment. By 1954 superior carcass yields were confirming that the cattle had promise, yet there was still little interest in them. However, Helen was still breeding her group, and in October 1959 Mervyn displayed the cattle to the public for the first time. Ten months later a pen of five of his bullocks fetched a record Australian price at the Melbourne saleyards, which sparked a visit to the Sutherland property by George Batten, a correspondent for *Country Life*, and George Simpson, the Beef Cattle Officer for the NSW Department of Agriculture. After various names for the breed had been discussed George Simpson suggested Murray Grey. The Murray Grey Beef Cattle Society was formed in the early 1960s, and the breed has since established a worldwide reputation.

The breed first arrived in the United Kingdom in 1975. It is a minority breed, with registrations remaining stable.

Murray Grey semen was first imported into New Zealand in 1969 and used on approved Angus base stock. The New Zealand Murray Grey Society was formed in 1970, and after the relaxation of restrictions on the importation of female cattle in the early 1970s a number of females were imported.

BREED FEATURES

Hardy and adaptable, growing well in climates ranging from cold high uplands to hot subtropical grasslands. Very docile temperament. Hybrid vigour makes it very useful for cross-breeding.

Breeding

Early maturing, with heifers able to be mated at 12 months of age. High fertility and easy calving, calf birth weight averaging 32 kg. Good mothering and milk production. Murray Greys have proved very successful in cross-breeding programmes, the cross with Charolais producing the Chargrey, and that with the Brahman creating the Australian breed called the Greyman.

Growth and carcass

Good growth rate and feed conversion results in early maturity at 20–24 months. The light to medium carcass has a very good yield and quality. Meat lean, evenly marbled and tender, and much sought after by Japanese restaurant and supermarket outlets.

Weight

Mature bull: Average weight 820 kg
Mature cow: Average weight 600 kg

APPEARANCE

A medium-sized muscular beef breed. Skin darkly pigmented, with coat colour ranging from dark grey to light silver. Naturally polled.

Above: (*left*) Murray Grey cow and calf. (*right*) Murray Grey bull.

Parthenais .... (B)

One of the oldest French breeds, derived from cattle in western France and Brittany. It was originally used for draught and milk production, its rich milk being ideal for making butter. A carefully designed breeding programme initiated in the late 19th century guided its development into a dual-purpose dairy-beef breed, and a herd book was established in 1893. It was at the height of its popularity during the early 20th century, but thereafter was increasingly replaced by single-purpose breeds, particularly the Friesian and the Charolais.

In 1973 the French Parthenais Breed Society established a programme with three main aims: to work towards the production of high-quality beef; to spread modern breeding and production techniques; and to ensure the high health status of the stock. Since then the breed has enjoyed a renaissance in France, and increasing interest is being shown in it in other countries.

The Parthenais was first imported into Britain in the late 1980s.

APPEARANCE

A comparatively long-bodied, medium-large, double-muscled breed. Coat colour can vary from fawn through to dark golden-brown, with paler-coloured areas on the legs, and circles around the eyes and muzzle.

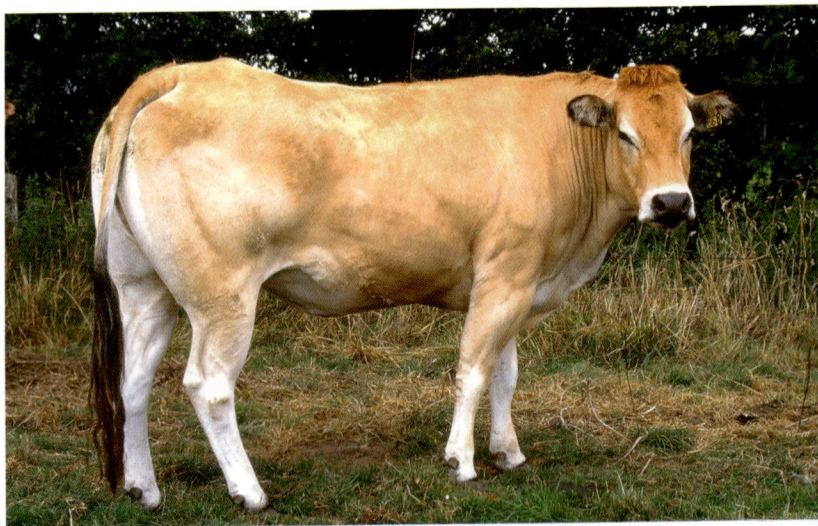

BREED FEATURES

Hardy and adaptable, thriving in all types of terrain and a wide variety of climates. Good disease resistance.

Breeding
High fertility, regular calving with about one-third of pure-bred births requiring minor assistance. Calf birth weight averages 42 kg.

Growth and carcass
Good growth, carcass killing out at over 67 per cent with high-quality lean, low-cholesterol meat.

Height and weight

Mature bull:	Average height at withers 147 cm	Average weight 1100 kg
Mature cow:	Average height at withers 142 cm	Average weight 800 kg

Left: Parthenais cow. *Above*: Parthenais bull.

Piemontese .... Ⓑ

In 1840 cattle native to the Piemont (Piedmont) region of north-west Italy were crossed with several imported European breeds, including the Iberian, and in 1887 a herd book was established. The herd book was closed four years later but in 1920 attempts to standardise the breed were revived, and a new herd book was set up in 1958.

The Piemontese has begun to make its mark outside Italy. In Holland in 1994 Piemontese accounted for 63 per cent of the beef bulls used in cross-breeding programmes. The resulting beef is marketed under the name 'Holland beef' in Germany and Italy. There is also increasing interest in Canada and the United States. The breed first entered Britain in 1988, and is gaining in popularity.

Australian stock originates principally from the importation of embryos and semen from Denmark and Canada.

The Piemontese first arrived in New Zealand in the 1980s.

APPEARANCE
A medium-sized long-backed beef breed, most individuals having characteristic double-muscling. Adult colour grey-white with black points. Naturally horned. Calves are fawn at birth, turning whiter as they grow. The fine skin is pigmented black, noticeable around the eyes, nose, tail tip and hooves.

Left: Piemontese bull.
Right: Piemontese cow and calf.

BREED FEATURES

Very docile. Hardy and adaptable to a variety of climates and altitudes. The double muscling is caused by a dominant gene and is therefore evident in cross-bred animals. The infusion of zebu blood provides some resistance to tick-borne diseases.

Breeding

Good fertility. Gestation about 7 days longer than the average for other breeds. Calf birth weight about 38 kg but may be much less. Few calving problems; calves are born slender and double muscling does not become evident until 3–6 weeks of age. Cows have good mothering ability and longevity. The breed is promoted particularly for cross-breeding programmes in commercial beef and dairy herds.

Growth and carcass

Calf birth weights can be quite low but their growth rate is very fast. Growing animals have a high feed conversion rate and daily weight gain. Comparatively small bones and thin skins contribute to a very high carcass yield that can exceed 70 per cent. Cross-breeds are suitable for both grain and grassland feeding. The meat is of high quality, lean, tender and tasty, with low cholesterol levels.

Height and weight

Mature bull:	Average height at withers 142 cm	Average weight 910 kg
Mature cow:	Average height at withers 132 cm	Average weight 595 kg

Pinzgauer (D)(B)

Descended from the red native cattle that roamed the region more than 1500 years ago, the Pinzgauer originated in Austria and Bavaria, and takes its name from Pinzgau, a province of Salzburg in Austria. It appeared for the first time in documents of the 17th century, and early in the 19th century herd books were already established, confirming that selective breeding had been continuing for some time. A breeding association was formed in the 1850s, and by 1890 three main types had been developed: the Molltal, Tyrol and Salzburg. In the early 20th century large numbers of Pinzgauer cattle were sent to South Africa, which now has the world's largest full-blooded herd. Originally used principally for draught and meat, the breed in Europe is now used mainly as a dual-purpose dairy-beef animal.

In New Zealand the Pinzgauer is recognised as a breed for the show ring, but at the time of writing breeders were awaiting confirmation that they could link up with Australia in order to classify their animals as pedigree. At that time there were only about 40 full-blooded Pinzgauers in New Zealand.

APPEARANCE
A medium-sized breed. Smooth, soft, medium-long hair and very pliable skin. A distinctive coat-colour pattern, light-to-deep chestnut red-brown with a white stripe commencing at the middle of the back, widening over the hindquarters, then down to the hocks and along to the brisket. A pigmented skin protects against ultraviolet rays.

BREED FEATURES

Quiet temperament. Adaptable to hot or cold climates. Good milk production results in rapidly growing and quick-maturing calves.

Breeding

Early maturity, females reaching puberty between 305 and 340 days. Good fertility. Easy calving, calf birth weight averaging 40 kg. Cows have a strong mothering instinct and good milk production. Cross-breeding produces calves with a high degree of hybrid vigour and strong ability to survive. Popular crosses are Pinzgauer × Simmental and Pinzgauer × Charolais.

Growth and carcass

Very good feed conversion, growth rate and carcass yield.

Height and weight

Mature bull:	Average height at withers 150 cm	Average weight 1100 kg
Mature cow:	Average height at withers 137 cm	Average weight 675 kg

Facing page: Pinzgauer cow and calf. *Above*: Pinzgauer bull.

Red Poll .... (D)(B)

The breed originated in England in the late 18th and early 19th centuries from the fusion of Norfolk Red cattle, chosen for their hardiness and beef production, and Suffolk Dun cattle, chosen for their milk production. Classes for the breed were first held at the Royal Show in 1862, although it was 1888 before a breed society was formed. It was originally a true dual-purpose animal, with cows producing 2500 litres of milk a season and steers killing out up to 300 kg carcass weight. Specialisation in milk production saw a decrease in the use of Red Polls and the breed then moved its emphasis to beef production, concentrating on functional cattle with a good-yielding carcass from vealers through to heavy-weights.

The Red Poll was imported into Australia in the 1850s. In 1898 three cows and a bull were imported from England into New Zealand to Otahuna, near Christchurch. Subsequent importation of live animals to both islands, and of semen from the United Kingdom, North America and Australia, is still occurring. The breed flourishes in England, Canada, the United States, Jamaica, Brazil, Uruguay, South Africa, Australia and New Zealand.

The New Zealand Red Poll Cattle Breeders' Association was formed in 1921 and has approximately 60 members, meeting annually for an AGM and herd tour of members in the appropriate district. Members regularly exhibit at shows, and the association has a large number of trophies competed for at the Royal Show.

APPEARANCE

A medium-sized dual-purpose breed. Colour a distinctive deep red. White hair is seen on the tail switch and occasionally on the udder. Naturally polled.

Left: Red Poll bull.

Right: Red Poll cow and calf.

BREED FEATURES

Very docile. Hardy, adaptable and a good forager, able to produce milk from poor grazing. Skin pigmentation helps to protect against ultraviolet light. Flexible marketing age supports veal production.

Breeding

Good fertility, easy calving, calf birth weight averaging 35–40 kg. Very good mothering tendencies, and cows will readily take a second calf for suckling. One of the longest-living breeds, with cows producing 15 or more calves. When used for cross-breeding the polled gene is dominant. The infusion of milking capacity makes it ideal to put a Red Poll cross female back to a terminal sire of a different breed.

Growth and carcass

Good milk production results in rapid weight gain by calf. Animals can be turned off anywhere from vealer stage through to heavyweight steers. Carcass yield high and of good quality with even fat cover. The tender fine-grained meat has led to success in carcass competitions in New Zealand and Australia.

Height and weight

Mature bull: Average height at withers 140 cm   Average weight 900 kg
Mature cow: Average height at withers 127 cm   Average weight 600 kg

Romagnola Ⓑ

A very old breed, descended from ancient Grey Steppe cattle with an infusion of *Bos t. indicus*, and named after the Romagna district of north-east Italy. The Romagnola is found principally in the provinces of Ravenna, Forli and Bologna, where the terrain varies from flat land of high quality to very steep ground which is difficult to traverse. In the upland region the vegetation becomes very dry and sparse during summer, yet cows nursing calves can still convert a frugal diet into milk and meat. To thrive, cattle must have the ability to walk over rough and very steep ground, and the breed is noted for its ability to walk vast distances. Particular attention has been paid to the feet and legs over many generations of breeding. On the more valuable lower land, adults and calves are permanently housed and fed on the by-products of the crops of the area.

The breed is well established in Italy, and also has a strong following in Europe and North America.

It was first imported into New Zealand in 1976.

APPEARANCE

Medium-sized, deep-bodied and strong-muscled with good rump and sturdy structure. Head carried high, hooded alert eyes, loose dewlap. Hooves strong and dark-pigmented. A feature of this breed is the straightness of the hind legs. Calves are born a honey fawn colour, changing to mature colour by 3–6 months of age. Adults have a basically white-grey coat, with dark hair extending over the shoulders and black points on the ears, nose and tail. The skin is thick, and darkly pigmented. The white coat colour is recessive when crossed with an unbroken colour breed. Bulls grow a long winter coat.

Left: Romagnola bull.

BREED FEATURES

A large powerful breed with rounded body and heavy legs. Semi tick-resistant. Excellent foraging ability. Highly adaptable, can stand extremes of heat and cold, and thrives in all zones from the tropics and arid areas to highly improved temperate climate pastures. Manageable intelligence. Bulls are being promoted as terminal sires in cross-breeding programmes with other beef breeds.

Breeding

Early maturity at about 13 months, high fertility. One of the few European breeds traditionally expected to calve under extensive grazing conditions. Calf birth weight averages 37 kg. Few birth problems and calf mortality is low. Calves grow rapidly, sometimes doubling their birth weight within two weeks. When crossed with other breeds these characteristics remain, so it is being promoted for use in cross-breeding programmes. In Australia Romagnola sires have produced good results when put over Brahmans, Angus and Murray Greys in large commercial enterprises, giving greater muscling. In New Zealand similar results have been obtained from crosses with Hereford, Shorthorn and Friesian × Hereford. Crossing with *Bos t. indicus* bloodlines results in a particularly large lift in bone and muscle development.

Growth and carcass

Beefing and conversion rates compare favourably with other widely established European breeds. Growth rate is rapid, and feed conversion highly efficient. Carcass yield good, carcass lean and smooth with even fat. Meat tender, rose-coloured and of fine texture.

Height and weight

Mature bull:	Average height at withers 158 cm	Average weight 1090 kg
Mature cow:	Average height at withers 144 cm	Average weight 680 kg

Right: Romagnola cow and calf.

Salers .... Ⓑ

This very old breed is named after the small medieval French town of Salers, which lies in a valley in the Cantal region of south-central France. It is a mountainous region with poor soil and a harsh climate, comparable to that of the Scottish Highlands. The breed was developed to forage during the summer months in the high mountain pastures, and to survive on grass and hay only. It was traditionally used for three purposes: draught, milk and meat. Some Devon (and possibly Highland) blood was introduced during the 19th century. Milk recording became compulsory in 1925, and weight recording commenced in 1962.

The export of Salers cattle was restricted until, in 1972, a French farmer was allowed to export a bull to his ranch in Quebec. Following further imports and grading-up programmes (including introduction of the poll gene) the breed spread throughout North America.

The first New Zealand imports were of live cattle originating from Canada and Australia, with embryos imported from Canada or the United Kingdom. In December 1986 a New Zealand Salers Development Group was set up, and in 1988 the New Zealand Salers Society was formed. Pure-bred registered herds within New Zealand range from small herds of 10–15 cows to large herds comprising up to 400 registered animals. All registered herds are recorded for estimated breeding values through ABRI, University of New England, in New South Wales.

APPEARANCE

A medium-sized, compact breed. Coat originally dark cherry red in colour, but black-coated animals are now appearing in pure-bred herds. In winter animals grow a thick, curly coat. Originally naturally horned, but polled animals are now becoming more common. Legs strong, with black hooves.

Left: Salers cow.

BREED FEATURES

Genetically a very pure breed, capable of introducing hybrid vigour when used in cross-breeding programmes. Very useful for cross-breeding in single-suckler herds. Extremely hardy, and a good forager that makes use of all available pasture and tolerates a wide temperature range. Can also tolerate long periods of intensive housing. Good longevity, about 25 per cent of cows reaching over 10 years of age. Brown pigmentation of skin and mucous membranes aids protection from the sun.

Breeding

Early puberty, with pure-bred heifers able to be mated at 15–17 months of age. High fertility, exceeding 80 per cent in heifers and 95 per cent in cows. Easy calving, calf long-bodied with fine shoulders and head. Calf birth weight averages 38 kg. Mean calving interval averages 374 days. Full maturity is reached comparatively late at 5–6 years of age.

Milk production

Traditionally the Salers cow was used in one lactation to produce both milk, which due to its high protein content is valuable for cheese production, and a veal calf ('broutard'). Although now used almost exclusively for beef production, the milk-producing qualities remain.

Growth and carcass

Growth rate is good, with calves gaining on average about 1 kg daily from birth to weaning. Carcass yields are high, and may reach 70 per cent or more. Little backfat, meat lean with good colour and marbling and excellent flavour.

Height and weight

Mature bull: Average height at withers 150 cm   Average weight 1100 kg
Mature cow: Average height at withers 140 cm   Average weight 750 kg

Right: Salers bull.

Santa Gertrudis .... (B)

This very successful cross between *Bos t. taurus* and *Bos t. indicus* was developed in Texas on the King Ranch, the largest in the United States. The object was to produce a breed that would perform well in the ranch's harsh, hot, dry conditions. In 1915 a large black Brahman × Shorthorn bull was crossed with Shorthorn cows. Female offspring were back-crossed with pure Shorthorn bulls. The resulting calves had potential, so a programme was begun to produce animals carrying 5/8 Shorthorn and 3/8 Brahman, selection being based on red colour and performance. In 1920 a bull calf was born with the required characteristics and blood proportions, and once mature this bull, called Monkey because of its playful disposition, was used as the sole stud sire from 1923 to 1932, during which period it sired 150 healthy bulls which were used to build the new breed. Therefore all Santa Gertrudis animals are descendants of Monkey. The breed was recognised in 1940, the King Ranch started selling Santa Gertrudis stock in 1950, and a breed society was formed a year later. A herd book was started in 1954. The breed is now found in about 50 countries worldwide. Breed quality and uniformity is aided by a rigid classification system.

The Santa Gertrudis Breeders' (New Zealand) Association was formed in 1969. Initially members operated a breeding-up programme using semen imported from Australia. The first pure-bred cattle born in New Zealand were classified in October 1973. The breed is now established throughout the country.

Left: Santa Gertrudis bull.

Right: Santa Gertrudis cows and calves.

BREED FEATURES

Adaptable to a wide range of climatic conditions. Excellent foragers. Females will 'babysit' calves while their mothers graze. High degree of heat, tick and bloat resistance. Exceptional longevity.

Breeding

Good fertility. Easy calving. Calf birth weight about 43 kg. Good mothering, good milk production with high milkfat content.

Growth and carcass

Efficient feed conversion on either natural pasture or in feedlots, high daily weight gain resulting in good weight-for-age. Good carcass yield. Fat evenly distributed.

Weight

Mature bull: Average weight 1050 kg
Mature cow: Average weight 700 kg

APPEARANCE

A large, long, strong-boned beef breed with heavily muscled hindquarters. Straight profile, slightly drooping ears, and unbroken dewlap from brisket to jaw. The male has a masculine crest immediately before the shoulders. Skin pigmented red, loose and slightly wrinkled in the neck folds. Hair short, coat colour deep cherry-red. Muzzle flesh-coloured. Naturally horned, horns pointing downwards and swept back. Polled strains now exist.

Shaver Beefblend ⓑ

A composite breed developed by Dr Donald Shaver, Senior, a Canadian who by the early 1960s had established an international reputation through the creation of genetically superior chickens, known as Shaver poultry, which revolutionised the poultry industry.

In 1961 he and his team commenced trials to produce a genetically superior beef breed, using combinations of 11 different existing breeds. By self-elimination the breed is now made up of nine breeds: Galloway, Highlander, Red Devon, South Devon, Lincoln Red, Gelbvieh, Salers, Blonde d'Aquitaine and Maine Anjou. The proportion of breeds is not made public, and animals are identified only by a name or number.

Within the Shaver breed there are several large families. This allows for skilful recombination, thereby maintaining the high level of heterosis (hybrid vigour) that is unique to the breed. None of the world's most populous breeds, such as the Holstein-Friesian, Hereford, Angus or Simmental, were used in the breed's development; consequently the use of Shaver bulls with these breeds expresses the maximum effects of heterosis. In effect, the Shaver can be used indefinitely in cross-breeding programmes with no reversion to any original breed type.

Shaver cattle continue to be selected for enhanced fertility and reproductive traits, improved growth rates and superior carcass quality, while retaining calving ease and good mothering ability.

BREED FEATURES

Enhanced fertility and reproductive traits, low birth weight, ease of calving and good mothering ability. Improved growth rates and superior carcass quality. Docile. Structurally sound.

Breeding

Early sexual maturity, high fertility, comparatively short gestation, easy calving, high calving percentage, good mothering. Calf birth weight averages 33 kg. Calves are rugged and vigorous, and adapt well to a broad range of conditions.

Growth and carcass

Comparative trials with other breeds have demonstrated very good growth rates, high weaning and final weights, and younger age at slaughter. In field trials in Mexico Shaver Beefblend × Brahman (Shabra) had an average daily weight gain nearly twice that of straight Brahman. Carcass yield and quality are high.

Weight

Mature bull: Average weight 1045 kg
Mature cow: Average weight 550–600 kg

Presently there are two studs in New Zealand, one in the central North Island and one in the South Island.

APPEARANCE

Males relatively large, females moderately sized. Naturally polled. Coat deep red.

Left: Shaver bull.
Right: Shaver heifer.

Shorthorn ⓓⓑ

The Shorthorn is one of the oldest British breeds, with a recorded history of well over 200 years. Its origins can be traced to the Tees River valley in north-east England, where the Teeswater beef breed evolved from a cross with a Dutch dairy type. An upgrading programme in Durham county produced cattle called Durhams, which became the pioneer bloodline of the Shorthorn. The Shorthorn was originally a dual-purpose animal, but by the early 19th century a breeder called Thomas Booth was selecting for beef qualities while another called Thomas Bates was selecting more for milking ability. In 1822 Volume 1 of the Coates's Herd Book was published, but it was not until 1958, with the publication of Volume 106, that the herd book was finally divided into beef and dairy sections.

In the United Kingdom after World War II both the Shorthorn breeds fell from favour, and efforts to improve the performance of the Dairy Shorthorn have since included the establishment of experimental programmes within the guidelines of the Shorthorn Society's Experimental Register. The result is that the breed has radically changed from that of 30 years ago, with smoother fleshing and substantially increased yields of milk and milkfat. In 1995 there were about 60,000 pedigree and 40,000 non-registered animals, and there are 3000–4000 new registrations each year.

Shorthorns were the first dairy cattle to enter New Zealand, two cows and a bull arriving from Australia in 1814. In both Australia and New Zealand the Dairy Shorthorn, now called the Milking Shorthorn, lost popularity as a milk-producer as first the Jersey and later the Holstein-Friesian took over.

Beef Shorthorn ⓑ

APPEARANCE

A medium-sized but muscular beef breed. The Australian Shorthorn is comparatively taller and heavier than other varieties. Colour red, roan or white. Naturally horned or polled.

Progress towards the 'traditional' Beef Shorthorn continued with the development of the Scotch Shorthorn, a beefy type of Shorthorn developed from 1837 onwards in Aberdeenshire, Scotland. During the 19th century the beef variety of Shorthorn became one of the most popular breeds in Europe, being crossed with local cattle to improve their performance. It was promoted as 'the great improver', and contributed to over 40 other breeds throughout the

BEEF SHORTHORN BREED FEATURES

Docile and easy to handle. Good maternal qualities. Early maturing. Good growth rate and carcass quality. Good for cross-breeding and as a terminal sire on dairy cows. Eye pigmentation reduces the risk of eye cancer in hot climates.

Breeding

Good fertility and early maturity. Easy calving with few problems, and very good mothering. Calf comparatively small, birth weight about 39 kg. The modern Beef Shorthorn is being used for cross-breeding: e.g. Beef Shorthorn cows mated with Continental sires, and Beef Shorthorn bulls mated with cows of various dairy breeds.

Milk production

The Beef Shorthorn was originally selected for beef production and animals from old lines have an unremarkable milk yield. However, cows from the improved lines are comparatively good milk producers, and their calves grow well and reach a very good weight by weaning.

Growth and carcass

The improved Beef Shorthorn is no longer overfat, but ranks among the leaner of the beef breeds. Calves grow particularly well pre-weaning, and produce a very good-quality carcass. The meat is well marbled with good flavour.

Weight

Mature bull: Average weight of old lines 800 kg
Mature cow: Average weight of old lines 500 kg

Right: Beef Shorthorn cow.

world. By the end of World War II it had become popular for the production of 'baby beef', animals which by today's standards were short, fat and dumpy.

From the 1950s onward, consumer demand for leaner meat combined with competition from other breeds led to a significant fall in numbers in the United Kingdom, and the horned variety of Beef Shorthorn from old British lines was eventually reduced to a few hundred breeding females. At the time of writing it is still classified as a rare breed, the latest survey showing approximately 600 females registered, the majority of herds being in Scotland and North Yorkshire.

In response to the change in consumer demand breeders have since worked to improve the Beef Shorthorn and also to develop newer types: for example, the Maine-Anjou has been used to produce animals of larger scale and height containing 63.5 per cent Shorthorn breeding. These efforts have led to renewed interest in the improved Beef Shorthorn breed.

Whitebred Shorthorn   (D)(B)

All-white animals have the same origin as the other Shorthorn beef breeds, and many white animals are registered in Coates's Herd Book. However, certain strains are registered separately in the United Kingdom. The White-bred Shorthorn is a special-purpose breed that was developed to cross with the Galloway in order to produce the Blue-Grey cross-bred suckler cow. As a type it is probably closer to the Dairy Shorthorn than the Beef Shorthorn.

New Zealand Shorthorn   (B)

The first Shorthorns in New Zealand were two cows and a bull that arrived from Australia in 1814. By 1842 many herds had been established and for a long time it was the main cattle breed, used for milk and meat production and for draught work. Here too, specific beef and dairy types were developed. Until the 1970s the Beef Shorthorn was the third most popular beef breed in the country, but numbers fell after the introduction of numerous European breeds for cross-breeding. Many New Zealand Beef Shorthorns now have an infusion of Australian bloodlines.

Right: New Zealand Shorthorn cow.

Facing page (*left*): Whitebred Shorthorn cow and calf.

Facing page (*right*): Polled Shorthorn cow.

Polled Shorthorn   (B)

Polled varieties of the Beef Shorthorn were developed in the late 19th century in America. One programme, started around 1881, mated polled native cows with horned Shorthorn bulls; it continued through five generations, breeding only from polled females. This was known as the single standard. Another, the double standard, used naturally occurring pure-bred Shorthorn polled mutant males and females to produce polled offspring, originally called Double Standard Durhams. In 1896 the rules in America were changed to ensure that only the progeny of 'recorded' Polled Durhams or 'registered' Shorthorns could be recorded and in 1919 the name was changed to Polled Shorthorn. In 1932 a polled breeding programme was established in Scotland, and in 1950 a Polled Shorthorn Society Herd of Great Britain was established. Further imports of polled Shorthorns from Australia and New Zealand resulted in a strong content of polled breeding in the United Kingdom. The Polled Shorthorn was introduced into New Zealand from America in 1936.

Dairy Shorthorn or Milking Shorthorn   Ⓓ Ⓑ

The early origins of the Dairy Shorthorn are the same as those of the Beef Shorthorn. A Dairy Shorthorn Association was formed in 1905, probably the first to begin a milk recording scheme in the United Kingdom, and was the first to make milk yield qualification a condition for showing. The Association was amalgamated with the Shorthorn Society in 1936. In New Zealand the Milking Shorthorn predominated until the 1920s when it was superseded by the Jersey. It is now a minority breed.

APPEARANCE
A medium-sized dairy breed. Colour red, roan or white. Originally horned, but polled strains now exist.

Above: Milking Shorthorn cow.
Right: Milking Shorthorn bull.

DAIRY SHORTHORN OR MILKING SHORTHORN BREED FEATURES

Docile and easily managed. Hardy and adaptable. Good productivity. Dual-purpose.

Breeding

Very good fertility, and has a record for lack of calving problems. Good longevity.

Growth and carcass

Mature Dairy Shorthorn cattle are as muscular as many beef breeds. In Australia cows are in demand as sucklers and for cross-breeding in commercial beef herds, while Dairy Shorthorn bulls find a ready market in commercial beef herds.

Milk production

In the United Kingdom the breed average in recorded herds in 1994 was 5600 kg of milk (many individual herds well over 6000 kg), 3.3 per cent protein and 3.84 per cent milkfat. This provides a protein/fat ratio of 0.86, the best ratio in terms of cash return of any breed in the country. The size of the fat globules makes the milk ideal for cheese production.

Weight

Mature bull: Average weight 920 kg
Mature cow: Average weight 625 kg

Simmental .... (B)

One of the oldest cattle breeds, which originated in the Simmen Valley in western Switzerland many centuries ago, and was much sought after for its rapid growth, outstanding production of milk, cheese and butter, and for use as a draught animal. The first herd register was established in Europe in 1806, where the national names for the breed are Fleckvieh (Germany), Pie Rouge (France), Pezzata Rossa Friulana (Italy) and Österreichisches Fleckvieh (Austria). The European Simmental is sometimes referred to as the Simmental-Fleckvieh, to distinguish it from the American and Australian varieties. The Simmental is numerically the second largest breed in the world (behind the Brahman), with a total population of about 42 million. About 20 million pure-bred Simmentals are registered throughout Europe; of these some 8–12 million are in Russia, while Simmentals account for 70 per cent of all the cattle in Austria. The breed also plays an important role in the Middle East, South Africa, North America and Australasia.

The Simmental entered Britain in 1970, and the British Simmental Society was founded that same year. The breed has made steady progress and 123,151 animals were registered at the beginning of 1995.

Above: Simmental cow.

BREED FEATURES

A medium–large, muscular breed with very good temperament. A well-proven breed of good conformation, fertility and maternal ability. Cows produce a good milk yield. Very good feed conversion, growth rate and carcass yield. An excellent breed for crossing with other beef and dairy breeds to improve growth rate and performance.

Breeding

Early maturity and high fertility. Females easy calving with very good maternal qualities. Calf birth weight averages 41 kg.

Growth and carcass

Cows have good milk production so their calves grow rapidly. Weight gain and feed conversion rates are high in comparison with most other beef breeds. Carcass yield is very good, and the meat is lean with little fat.

Height and weight

Mature bull: Average height at withers 152 cm   Average weight 1250 kg
Mature cow: Average height at withers 138 cm   Average weight 750 kg

BLACK SIMMENTAL BREED FEATURES

Particularly popular for cross-breeding programmes requiring the black coat colour and polled factor. Crossed with the Hereford the white-faced progeny have either black or grey coats. There is strong market demand, particularly in Japan, for black- or grey-coated cattle.

RED SIMMENTAL BREED FEATURES

Particularly useful for cross-breeding programmes in which a red coat colour is to be retained.

The Simmental breed was introduced into New Zealand in 1971 through semen importation from England and Scotland, out of donor sires originating from German stock. The Simmental Society of New Zealand was formed just prior to that in order to oversee a breeding-up programme. In 1972 a small number of pure-bred cattle were imported from England, and the first calves from the breeding programme were born. The breed is steadily gaining in popularity, especially for cross-breeding.

APPEARANCE

A large, well-muscled breed. In Germany and Austria many Fleckvieh are of the more milky type, but in Germany some are more thickset like the old type of beef Simmental. Hair soft. Body colour ranges from pale yellow or straw-coloured through to deep red. There are white markings on the head, underside of the brisket and belly, legs and tail, and sometimes on the body, particularly behind the shoulders and on the flanks. There is often dark pigmentation around the eyes, and many animals have large eye patches. Naturally horned. Some polled animals are being bred. There are also Red Simmentals and Black Simmentals, for which separate registers are kept.

Black Simmental   (B)

The same origins as the Simmental, but later derived from grading-up with black-coated breeds, particularly the Aberdeen-Angus, Galloway and Holstein-Friesian. Pure-bred Black Simmentals contain at least 15/16 Simmental content.

APPEARANCE

Black or grey coat, with white face and head, often with black eye patches. Naturally polled.

Red Simmental (B)

Evolved from the Simmental and Black Simmental, animals being selected for having more red colouring on the head, body and legs. Many are solid red, with the red gene being dominant.

APPEARANCE
Body colour chestnut red to dark red. Red eye pigmentation.

Above: Red Simmental cow and calf.

Left: Red Simmental bull.

Right: Black Simmental bull.

South Devon .... (B)

Like many European breeds the origin of the South Devon is unclear, but it may well owe part of its genetic make-up to the large red cattle of Normandy, which were introduced into Britain from France at the time of the Norman invasion in AD 1066. Zebu (*Bos t. indicus*) blood may have been introduced when cattle were brought from India to Plymouth on trading ships. Whatever the origin, these distinctive copper-red cattle were firmly established in southwest England by the 16th century, and bullocks of the breed were used as draught animals. The South Devon Herd Book Society of England was formed in 1891. In those days the breed was valued for its high milkfat production, producing much of the famous clotted cream used in Devonshire teas. Less attention was paid to its value as a meat producer until the breed began to achieve worldwide prominence during the 1960s, when weight-gain recording became popular and the breed's excellent performance became recognised. South Devons can now be found on many continents in a range of climatic conditions.

The South Devon arrived in New Zealand in 1969.

APPEARANCE
The largest of the English beef breeds, with a distinctive, even, rich copper or medium-red coat colour. Naturally horned, but polled animals also exist.

Left: South Devon bull.

Right: South Devon cow and calf.

BREED FEATURES

Hardy, retaining hybrid vigour, an excellent beef and milk producer. Versatile and able to forage and thrive under a wide variety of conditions, it is considered to be at its best under range conditions. An excellent temperament has earned the breed the nickname of 'gentle giant'. Red skin pigmentation helps to protect against the sun and eye cancer, and it has been successfully bred and reared in the tropics although in such climates the South Devon, like many breeds, benefits from a further infusion of *Bos t. indicus* blood.

Breeding

Early puberty with high fertility. A wide pelvis ensures easy calving. Calf birth weight about 43 kg, calves long-bodied and vigorous. Good mothering ability, and good milk production. A consistent breeder with a long life span. Considered ideal for cross-breeding with both beef and dairy breeds to enhance productivity.

Growth and carcass

Very good growth rates and feed conversion. Meat of very good quality, lean and tender.

Height and weight

Mature bull:	Average height at withers 155 cm	Average weight 1250 kg
Mature cow:	Average height at withers 135 cm	Average weight 750 kg

Sussex

One of the oldest breeds in England, whose ancestors were roaming the forests of Kent and Sussex well over 1000 years ago. A Mr John Ellman of Glynde, who also helped to develop the Southdown sheep, is credited with starting to improve the Sussex breed in the 18th century, selecting them for working and for beef. The Royal Agricultural Society recognised the breed in 1851, and a herd book was commenced in 1874. Sussex cattle were used as working animals right through the early part of the 20th century, but their value as meat producers and the advent of mechanisation gradually led to them being bred for beef.

Sussex have been exported to a number of countries, the most impact being in southern Africa where they have been in demand for over 70 years because of their inherent resistance to heat stress, combined with their tolerance of cold.

The breed entered Australia in 1915, and proved capable of handling hot, dry subtropical conditions.

The Sussex first entered New Zealand in the early 1970s, and a breed society was formed soon after. However there are only a few herds in the country, resulting from both imported pure-bred stock and graded-up animals.

Above: Sussex bull. *Right*: Sussex cow.

BREED FEATURES

Very docile. Adaptable, with good foraging ability and tolerance to drought.

Breeding

Good fertility, easy calving with very low calf mortality. Calf birth weight about 35 kg. Average calving interval one of the shortest of the beef breeds. Bulls recommended as terminal sires for cross-breeding.

Growth and carcass

Early maturing. Good feed conversion. Meat of high quality, texture and colour. To reduce depth of backfat and increase size some grading-up programmes use either Limousin or Salers.

Height and weight

Mature bull:	Average height at withers 145 cm	Average weight 950 kg
Mature cow:	Average height at withers 135 cm	Average weight 700 kg

APPEARANCE

A medium-sized beef breed, with firm even fleshing, non-protruding shoulders, well sprung ribs, straight back, broad loin and muscular hindquarters. Moderately thick skin, soft to the touch. Coat a dark mahogany red, extremely smooth in summer. Traditionally horned, a polled strain has been bred since 1950 and polled animals have been registered since 1979.

Texas Longhorn .... Ⓑ

In 1493, on his second voyage to the New World, Christopher Columbus took some Spanish cattle to Santa Domingo in South America. In 1521 a group of these cattle were taken from Santa Domingo to Mexico, where their numbers multiplied. In 1690, 200 cattle from the Mexican ranges were driven north to a mission near the Sabine River, an area which later became part of the State of Texas. Over the next hundred years numbers increased rapidly, not so much through the efforts of ranchers but as a result of natural breeding among the many animals that escaped or were left behind after ranches failed. By 1790 millions of cattle roamed the state, most of them unbranded. During the next 25 years, 10 million cattle were trailed northward to the mid-western grasslands, and even more were shipped eastward. They were noted for their endurance and resistance to disease, being able to withstand long trail drives, go long distances without water, swim rivers, survive in hot desert sun and winter snow, and fend for themselves. By the beginning of the 20th century, however, diminishing demand for Longhorns and an increasing demand for European breeds had caused a rapid decline in numbers, and by the late 1920s there were only a few hundred pure-bred cattle left. It was not until the 1950s that interest in the breed was revived, but since then numbers have increased. At the time of writing more than 100,000 pure-bred Texas Longhorns are registered with the Texas Longhorn Breeders' Association of America, and the breed is arousing interest in Australasia for cross-breeding with yearling heifers of both dairy and beef breeds.

The Texas Longhorn was first introduced into New Zealand in 1989, where it is a minority breed.

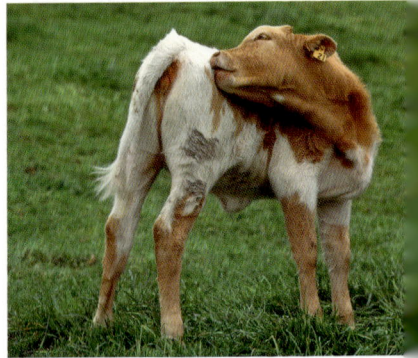

BREED FEATURES

Remarkably hardy and adaptable to a wide variety of climatic conditions. Comparatively disease-resistant. The breed offers genetic diversity, and animals are relatively long-lived. Docile.

Breeding

High fertility, easy calving, good mothering. Calf at birth weighs around 25–30 kg, and is vigorous. Potential for cross-breeding with yearling heifers of other breeds.

Growth and carcass

Not fast-growing, but produces good weaning weights. Very lean carcass.

Weight

Mature bull: Average weight 750 kg
Mature cow: Average weight 400 kg

APPEARANCE

Large-framed with characteristically long, upward-curving horns. Shoulders and pelvis noticeably narrow, allowing very easy calving. There are 225 possible colour combinations.

Far left: Texas Longhorn cow. *Left*: Texas Longhorn calf. *Above*: Texas Longhorn bull.

Wagyu (Mishima; Japanese Black) .... Ⓑ

The Wagyu is the national beef breed of Japan, accounting for about 85 per cent of that country's cattle population. It is derived from the original native cattle of Japan, whose ancestors were of the type known as Turano-Mongolian. Crossing with European breeds commenced in 1872 with the introduction of Ayrshire, Simmental, Swiss Brown and possibly Shorthorn blood, but the export of Wagyu cattle from Japan was prohibited for more than a century. Since 1910 Wagyu cattle have been closed to outside bloodlines. The breed became prized by the Japanese for its outstanding carcass quality and palatability, the result of a high degree of marbling in the meat.

In 1976 four Wagyu bulls were exported to the United States, and following semen collection they were purchased by Wagyu Breeders Inc. All North American and Australasian bloodlines were derived from these initial importations. In 1993 two full-blooded males and, for the first time ever, three full-blooded female cattle landed in the United States. Thirty-five more full-blooded Japanese cattle arrived in 1994. These importations resulted in dramatic improvements in both the quality and consistency of the cross-bred progeny.

In New Zealand the first Wagyu was born in 1991 as a result of embryo transplant. Six months later a pure-bred heifer was imported from Australia. Cross-breeding programmes are well established, most using Aberdeen-Angus.

BREED FEATURES

Docile, and capable of withstanding a wide variety of climatic conditions. Wagyu meat has a particularly high degree of marbling, which is highly heritable.

Breeding

Early maturity. Very easy calving. Calf birth weight comparatively low, at 33 kg.

Growth and carcass

In Japan all beef is grain-finished, and animals routinely reach 550 kg at 26–28 months of age. The meat is subjected to a sophisticated grading system on a scale of 1–5. Carcasses are also graded for yield from A (highest) to C. Wagyu normally grade higher than B3. In Canada and the United States Wagyu meat is marketed under the trademark of Kobe Beef.

Height and weight

Mature bull:	Average height at withers 139 cm	Average weight 750 kg
Mature cow:	Average height at withers 125 cm	Average weight 425 kg

APPEARANCE

A compact, short-necked, fine-boned breed with slim legs. The cylindrical body has heavy front shoulders, making it somewhat wedge-shaped. The Wagyu is slightly smaller in size than the Aberdeen-Angus, and naturally horned. The coat is particularly soft. There are two strains identifiable by colour: dark red-brown, indicating the Simmental influence, and black (a closed breed and the preferred colour).

Grain-fed animals weigh more.

Left: Wagyu cow.
Right: Wagyu bull.

Welsh Black .... (B)

The breed is descended from cattle that were domesticated in Britain at the time of the Roman Conquest in 55 BC. During the Saxon invasions between AD 827 and 1066, the retreating Britons took animals with them into the Welsh mountains. References to Welsh cattle can be found in old Welsh literature of the Middle Ages. The present breed society was formed in Wales in 1904. The breed is now established throughout the United Kingdom and in many other countries. At the time of writing there were 250 UK-registered herds, with about 2000 calves being registered annually.

Introduction of the breed into New Zealand was by artificial insemination in 1970, the importation of two pure-bred bulls in 1973, and a later importation of seven in-calf heifers. At the time of writing all remaining Welsh cattle in New Zealand had either been bred by artificial insemination or were the progeny of those early imports.

APPEARANCE

A medium-sized breed with a long, wide, deep body. Naturally horned, but a polled strain has been developed. Winter coat fairly long, thick, mossy, black or rusty black in colour, replaced in summer or in hot climates by a smooth glossy black coat. A red factor is present in some individuals, and red-factor cattle are now eligible for registration in New Zealand.

Below: Welsh Black bulls. *Facing page*: Welsh Black cow and calf.

BREED FEATURES

A true hill-country breed, originally dual-purpose and bred to withstand harsh outdoor environments and unimproved pasture. Now regarded as an easily managed quality suckler beef producer. The breed has retained the willingness to forage, and the ability to convert rough feed into weight gain. Docility makes for easy handling. Females have a prime role as a suckler cow. Males are prime beef producers.

Breeding

Bulls virile, cows highly fertile. Easy calving. Newborn calf weight approximately 33 kg. Cows have good mothering ability and a long, even lactation, averaging about 9 months, with a productive life of 15–20 years. Because of the wide pelvic area in females, and their stamina and milking ability, the breed is considered ideal for cross-breeding with beef breeds such as the Charolais and Maine-Anjou, and suitable for dairy beef production from dairy breeds such as the Jersey, Friesian and Ayrshire.

Growth and carcass

Animals fatten readily on good pasture, but do not become overfat. They show good feed conversion and weight gain. Carcass yields are high, and tests have shown the Welsh Black to be the leanest of the British breeds. The meat is fine-grained and lightly marbled, with good flavour.

Height and weight

Mature bull:	Average height at withers 147 cm	Average weight 950 kg
Mature cow:	Average height at withers 132 cm	Average weight 575 kg

Glossary ....

Agistment	The grazing of animals on land controlled by another person; usually for the payment of a fee.
Artificial insemination	The introduction of semen into a cow by a human operator via a pipette
Butterfat	*See* milkfat
Calving index	The interval between successive calvings
Carcass yield	The proportion of useable meat, as a percentage of total carcass weight
Cross-breeding	Mating two different breeds to produce a mixed breed
Curd	A soft substance formed by the coagulation of milk; used in making cheese or eaten as a food
Dewlap	Fold of skin running from under the neck down to the brisket (chest)
Dominant gene	A gene that exerts its effect even when not present as a pair. *See also* recessive gene
Donor cow	A cow from which artificially fertilised ova (embryos) are collected
Embryo transfer	The technique of transferring embryos from one cow to another
Eye cancer	Cancerous condition of the eye caused by reaction to ultra-violet light
Feed conversion	The proportion of food that is converted into weight gain or milk production
Feed lot	An area where cattle are yarded in groups and fed supplementary feeds
First cross	The offspring of a mating between two different breeds
Fold	A collection of livestock.
Gestation	The length of time from successful mating to birth of the calf
Grading up	A breeding programme to improve the quality of a group of animals
Heifer	A female that has not yet given birth to a calf
Heterosis	A beneficial effect resulting from a combination of different genetic profiles
Hybrid vigour	See heterosis

Introgression	The introduction into a given breed of genes that are not in its normal profile
Longevity	Life span
Marbling	The layering of fat between muscle fibres
Milkfat	The fat globules found in milk
Milk protein	The protein content of milk
Pizzle	The penis and its surrounding sheath
Poll, polled	Without horns
Puberty	Sexual maturity
Recessive gene	A gene that only exerts its effect when present as a pair. *See also* dominant gene
Recipient cow	A cow into which an artificially fertilised ova (egg), often of another breed, is introduced
Rotational grazing	Moving animals at regular intervals from a grazed paddock to an ungrazed paddock until all available land has been used and the cycle recommences
Steer	Castrated male
Stocking rate	The number of animals that can be kept on a given area of land
Suckler cow	A cow used for rearing one or more calves that are not necessarily her own
Terminal sire	A bull whose offspring are destined for slaughter, not herd replacement
Withers	The highest part of the back of a beast, behind the neck between the shoulders

Cattle breed categories ....

DAIRY BREEDS

Ayrshire
Brown Swiss (Braunvieh)
Guernsey
Holstein-Friesian, Holstein
Jersey

Meuse Rhine Issel
Dairy Shorthorn
Pinzgauer
Red Poll
Dexter

BEEF BREEDS

Angus or Aberdeen-Angus
   Red Angus
Aubrac
Australian Lowline
Belgian Blue
Belted Galloway
Blonde d'Aquitaine
Brahman
British White
Brown Swiss (Braunvieh)
Charolais
Chianina
Devon (Red Devon)
Dexter
Galloway
Gelbvieh
Hereford
   Poll Hereford
   Miniature Hereford
Highland
Limousin
Longhorn
Luing
Maine-Anjou
   Black Maine

Mandalong Special
Marchigiana
Meuse Rhine Issel
Murray Grey
Parthenais
Piemontese
Pinzgauer
Red Poll
Romagnola
Salers
Santa Gertrudis
Shaver Beefblend
Shorthorn
   Beef Shorthorn
   Whitebred Shorthorn
   New Zealand Shorthorn
   Poll Shorthorn
Simmental
   Black Simmental
   Red Simmental
South Devon
Sussex
Texas Longhorn
Wagyu
Welsh Black

Index

Page numbers in **bold** indicate the main information about the breed.